SpringerBriefs in Applied Sciences and Technology

Manufacturing and Surface Engineering

Series Editor

Joao Paulo Davim ⓘ, Department of Mechanical Engineering, University of Aveiro, Aveiro, Portugal

This series fosters information exchange and discussion on all aspects of manufacturing and surface engineering for modern industry. This series focuses on manufacturing with emphasis in machining and forming technologies, including traditional machining (turning, milling, drilling, etc.), non-traditional machining (EDM, USM, LAM, etc.), abrasive machining, hard part machining, high speed machining, high efficiency machining, micromachining, internet-based machining, metal casting, joining, powder metallurgy, extrusion, forging, rolling, drawing, sheet metal forming, microforming, hydroforming, thermoforming, incremental forming, plastics/composites processing, ceramic processing, hybrid processes (thermal, plasma, chemical and electrical energy assisted methods), etc. The manufacturability of all materials will be considered, including metals, polymers, ceramics, composites, biomaterials, nanomaterials, etc. The series covers the full range of surface engineering aspects such as surface metrology, surface integrity, contact mechanics, friction and wear, lubrication and lubricants, coatings an surface treatments, multiscale tribology including biomedical systems and manufacturing processes. Moreover, the series covers the computational methods and optimization techniques applied in manufacturing and surface engineering. Contributions to this book series are welcome on all subjects of manufacturing and surface engineering. Especially welcome are books that pioneer new research directions, raise new questions and new possibilities, or examine old problems from a new angle. To submit a proposal or request further information, please contact Dr. Mayra Castro, Publishing Editor Applied Sciences, via mayra.castro@springer.com or Professor J. Paulo Davim, Book Series Editor, via pdavim@ua.pt

More information about this subseries at http://www.springer.com/series/10623

Swagata Samanta · Pallab Banerji ·
Pranabendu Ganguly

Photonic Waveguide Components on Silicon Substrate

Modeling and Experiments

 Springer

Swagata Samanta
University of Glasgow
Glasgow, UK

Pallab Banerji
Indian Institute of Technology Kharagpur
Kharagpur, West Bengal, India

Pranabendu Ganguly
Indian Institute of Technology Kharagpur
Kharagpur, West Bengal, India

ISSN 2191-530X   ISSN 2191-5318   (electronic)
SpringerBriefs in Applied Sciences and Technology
ISSN 2365-8223   ISSN 2365-8231   (electronic)
Manufacturing and Surface Engineering
ISBN 978-981-15-1310-7   ISBN 978-981-15-1311-4   (eBook)
https://doi.org/10.1007/978-981-15-1311-4

This Springer imprint is published by the registered company Springer Nature Singapore Pte Ltd.
The registered company address is: 152 Beach Road, #21-01/04 Gateway East, Singapore 189721, Singapore

Dedicated to our parents and family

Preface

This monograph intends to provide a clear view of the theoretical and ongoing experimental methods for the fabrication of silicon and SU-8 polymer waveguides and some structures based on these waveguides for optical integrated circuit applications. The work is intended for researchers, scientists, and fabrication engineers working in the field of integrated optics, optical communications, laser technology, and optical lithography for device manufacturing.

All the work discussed in this monograph has been carried out using the research facilities of the Advanced Technology Development Centre (ATDC) and Central Research Facility (CRF) of the Indian Institute of Technology Kharagpur (IITKGP). The authors would like to acknowledge all the members of the centre who have directly or indirectly supported in the completion of the work and especially Prof. S. K. Lahiri for his thought-provoking ideas through discussion sessions. A big thank to Prof. Sakellaris Mailis and Prof. Srinivas Talabattula for their suggestions in improving the monograph content. The editors and publishers of respective journal articles (Elsevier GmbH, Elsevier B.V., Cambridge University Press, and IOP Publishing Ltd.) are also highly acknowledged for granting permission to reuse data content, figures, and tables.

Glasgow, UK   Swagata Samanta
Kharagpur, India   Pallab Banerji
Kharagpur, India   Pranabendu Ganguly

Contents

About the Authors

Dr. Swagata Samanta received her Ph.D from the Advanced Technology Development Centre (ATDC), Indian Institute of Technology Kharagpur (IIT-KGP) in 2018. She continued her research as a postdoctoral fellow at the Centre for Nano Science and Engineering (CeNSE), Indian Institute of Science (IISc) Bangalore. Presently, she is a postdoctoral research assistant in the School of Engineering, University of Glasgow, Scotland, UK. Her research interests include novel on-chip nanophotonic and nanoelectronic devices, integrated optics, VLSI systems, image processing, and artificial intelligence.

Dr. Pallab Banerji obtained his Ph.D from Jadavpur University, Kolkata, India. Presently, he is serving as a Professor and Head of the Materials Science Centre, Indian Institute of Technology Kharagpur. His major research areas are low dimensional semiconductors: structures and devices, photonics, thermoelectrics, and compound semiconductors. He has published about 120 research papers in international journals and guided more than 15 doctoral students.

Dr. Pranabendu Ganguly received his Ph.D in 2000 from the Indian Institute of Technology Kharagpur in integrated optics. Currently, he is working as senior scientific officer at the Advanced Technology Development Centre (ATDC), IIT Kharagpur. His recent areas of interest include micro- and nano-photonic devices, and guided wave optics. Dr. Ganguly has published 95 research papers in national and international journals, and conference proceedings. He is a Fellow of the Optical Society of India.

Abbreviations

1-D	One-dimensional
2-D	Two-dimensional
AFM	Atomic force microscopy
BCB	Benzocyclobutene
BPM	Beam propagation method
CMOS	Complementary metal–oxide–semiconductor
CMT	Coupled mode theory
DI	Deionized
DWDM	Dense wavelength division multiplexing
EBL	Electron beam lithography
EIM	Effective index method
EIMM	Effective index-based matrix method
FDTD	Finite-difference time-domain
FEM	Finite element method
FESEM	Field-emission scanning electron microscope
FIB	Focused ion beam
FSR	Free spectral range
FWHM	Full width at half maximum
ICP	Inductively coupled plasma
IPA	Isopropyl alcohol
LSI	Large-scale integration
MRR	Micro-ring resonator
NoC	Network on a chip
OIC	Optical integrated circuit
PDMS	Polydimethylsiloxane
PECVD	Plasma-enhanced chemical vapor deposition
Q factor	Quality factor
RIE	Reactive ion etching
rms	Root-mean-square
Si	Silicon

SiO_2	Silicon dioxide
SNR	Signal-to-noise ratio
SOI	Silicon on insulator
TE	Transverse electric
TM	Transverse magnetic
TMM	Transfer matrix method
WDM	Wavelength division multiplexing

Symbols

t	Thickness of silicon dioxide
h	Wire core thickness
h_s	Thickness of slab
H	Thickness of rib
w	Wire width
w_1	Rib width
n	Refractive index
n_1	Refractive index of core (silicon/SU-8)
n_b	Refractive index of silicon dioxide
n_c	Refractive index of cover
n_0	Refractive index of hypothetical layer
$n_{eff,m}$	Effective index of guided mode
n_g	Group index
$n_{eff,slab}$	Effective index of slab
$n_{eff,0}$	Effective index of fundamental mode of rib waveguide
n_{effo}	Effective refractive index or mode index of fundamental mode of waveguide
n_s	Substrate refractive index
n_{eq}	Equivalent refractive index
n_i	Refractive index of ith layer
n_{rib}	Refractive index of rib
n_{slab}	Refractive index of slab
d_i	Thickness of ith step refractive index layer
d_1	Distance between the added layer and guided region
Δn	Refractive index contrast
N	Number of step refractive index layers
ψ	Field amplitude
E_i	Electric field
m	Mode number
k_0	Free space propagation constant

β	Propagation constant
β_s	Symmetric propagation constant
β_a	Antisymmetric propagation constant
λ	Wavelength
λ_0	Resonating wavelength
y	Distance along the propagation direction
θ	Angle of incidence
δ	Detuning parameter
E_i^+	Electric field amplitude of transmitted propagating waves
E_i^-	Electric field amplitude of reflected propagating wave
e_i^+	Unit vector along the electric field direction (downward)
e_i^-	Unit vector along the electric field direction (upward)
r_i	Fresnel amplitude reflection coefficient at the ith interface
t_i	Fresnel amplitude transmission coefficient at the ith interface
S_0	Minimum separation
S_1	Maximum separation
g_0	Input optical wave
g_1	Output optical wave in waveguide 1
g_2	Output optical wave in waveguide 2
R	Radius of curvature of waveguide
g	Separation
κ	Coupling coefficient
C	Overall coupling coefficient
κ^2	Power coupling coefficient between the bus and ring waveguide
κ_p^2	Propagation power loss coefficient per round trip in the ring resonator
L_c	Coupling length
L	Length
φ	Phase
Δ	Phase difference
AR	Amplitude ratio
Γ	Full width at half maximum
r	Bending radius
BL	Bending loss
l	Circumference
$T_{through}$	Through port power transmission
T_{drop}	Drop port power transmission
ER	Extinction ratio
δT	Temperature deviation
$A_i(0)$	Amplitude at input end of three-coupled waveguides
$A_i(L)$	Amplitude at output end of three-coupled waveguides
Π	Excess loss
P_{in}	Fiber output–waveguide input
P_{out}	Waveguide output
D	Lateral offset between the two parallel waveguides

L_{bend}	Transition length in the longitudinal direction
T	Transition loss
∂_m	Modal offset between two arc bends
ρ	Half-width of waveguide
a_x	Spot size

Chapter 1
Introduction to Waveguides

1.1 Fundamentals and Background

Optical waveguides are structures which guide waves (flow of optical energy) in the optical spectrum. These can be broadly categorized into planar and non-planar waveguides; non-planar waveguides can be further classified according to geometry, mode structure, refractive index distribution, and material [1]. Figure 1.1 illustrates different types of optical waveguides.

Planar or slab waveguides are one-dimensional waveguide structures which confine light (waves) only in one transverse direction. Theoretically, these are infinite (practically not) in the direction parallel to the interface; however, if the interface size is extremely large with respect to the depth of the layer, then the approximation will be too good. Non-planar waveguides have confinement in both transverse directions. According to geometry, these can be divided into channel waveguides and optical fibers; channel waveguides can be further divided into strip, strip-loaded,

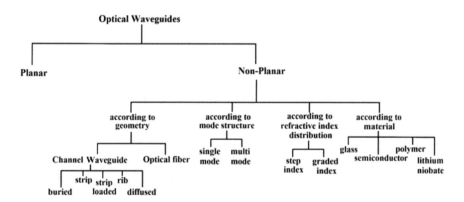

Fig. 1.1 Types of optical waveguides

buried, rib, and diffused waveguides. Wire or strip waveguide is the one which has strip or ridge on top of its planar structure (i.e., slab). A strip loaded on a planar waveguide forms the strip-loaded waveguide. If a high-index core is buried within a low-index surrounding medium, buried channel waveguide is formed. The structure of rib waveguide is similar to a wire waveguide, only difference is that the strip is of same refractive index as the high-index planar layer situated below it, and is a part of the waveguide core. Diffused waveguide can be formed if a high-index region is created within the substrate by process of diffusion of dopants. Optical fiber is made up of dielectric material which is surrounded by a lower refractive index dielectric material. According to mode structure, non-planar waveguide may be either single-moded or multi-moded in nature. The waveguide supporting only one mode (i.e., fundamental mode) is known as single-mode waveguide, whereas waveguide which supports higher order modes is multi-moded. Waveguides can be step-indexed or graded-indexed according to refractive index distribution. If the change in refractive index profile between core and cladding is abrupt, it is a step-index waveguide while the index profile that changes gradually is a graded-index waveguide. This catego-rization can also be made according to materials by which they are made of like glass, semiconductor, lithium niobate, and polymer [1–4].

Among the semiconductors, silicon (Si) has been the dominating material in elec-tronic industry for the last few decades; this dominance has also been extended into the field of microphotonics. Si waveguide forms the basic building block of all Si optical integrated circuits (OICs). It is optically transparent at telecommunication wavelengths ranging between 1.3 and 1.6 μm, so Si film of silicon-on-insulator (SOI) substrate can be used to fabricate low-loss optical waveguides; also, it is com-patible with the complementary metal–oxide–semiconductor (CMOS) technology. The high-index contrast of Si with silicon dioxide (SiO_2) or air benefits in strong light confinement, which makes it possible to fabricate very compact optical devices with bent waveguide radii in the order of a few micron, and functional waveguide elements like micro-ring resonator of ten to few hundred microns; thus, large-scale integration (LSI) of many functional elements on a single chip can be obtained. However, with these huge merits, there are some challenges for Si. Being an indirect bandgap material, Si is inefficient as a light emitter. Also, being centrosymmetric, it lacks electro-optic and nonlinear optic properties, thus making electro-optic Si modulators difficult to realize [5–7].

On the other hand, among the well-known polymers for photonic integrated cir-cuits, SU-8 is the most popular one. It is a high-contrast epoxy-based (consists of eight epoxy groups) negative resist. Optical transparency in the visible region as well as telecommunication wavelengths of 1.3–1.6 μm makes it a multifaceted material for OICs [8–10]. SU-8 was originally developed by IBM in Yorktown in the late 1980s and was designed for the fabrication of microstructures of high aspect ratios. Currently, it is commercially available from two companies, viz. Microchem Corpo-ration (Westborough MA, USA) and Gersteltec Sarl (Pully, Switzerland) in various formulations. Initially, it was developed as a thick-film resist for the patterning of molds for electroplating in the LIGA process, but very soon it became a popular material in other areas of microfabrication including microfluidics and photonics

[11–14]. Additionally, this polymer is chemical resistant and to some extent bio-compatible, which is of great advantage in designing lab-on-a-chip and microfluidic devices [15–18]. For on-chip optical interconnects, this low-cost SU-8 waveguide is useful rather than Si waveguide as the refractive index of SU-8 being lower than Si, thus yielding faster distribution of optical signals at different nodes of the chip [19].

The research in waveguiding started with planar waveguides and continued with large rib waveguides. Thereafter, there had been a trend to reduce the waveguide dimensions, and as a result, both small cross-sectional rib and wire waveguides were realized. Reducing the waveguide dimensions helps in achieving single-mode waveguides, the benefit of which is strong confinement and low optical loss. Also, the single-mode waveguides find their utility in interconnecting components on OICs and in preventing unwanted crosstalk between different waveguides on the substrate.

A thorough study on optical waveguides and waveguide-based devices has already been made by several research groups. In this monograph, some studies on rib and wire silicon and SU-8 waveguides and some structures (such as micro-ring resonator, photonic crystal, three-waveguide power splitter) were undertaken. In the following subsections, a brief review on these topics is made.

1.1.1 Rib and Wire Waveguides

The pioneering work of Soref et al. in [20] led to the beginning of research in single-moded silicon waveguides. Soref applied the concept of Petermann [21] (who showed that rib waveguide of large cross-sectional area which can be comparable to optical wavelength can behave as single-mode structure) and is the first to propose the expression for single-mode large cross-sectional rib waveguides (LCRW)—this is known to be Soref's condition by his name. Later, Pogossian et al. [22] proposed a modified formula of this Soref's condition and compared the theoretical results using effective index method with experimental data for single-mode semiconductor LCRW, which is claimed to be stronger condition for single-mode design purposes. Powell [23] investigated the conditions for both vertical and slanted trapezoidal-walled rib structures by beam propagation method, thereby pointing the inadequacy of using the effective index method for the prediction of LCRW in the cut-off regions. Thereafter, research interest moved to smaller device dimensions for better device performance and cost-efficiency. Vivien et al. [24] in 2002 reported on numerical simulations of single-mode as well as polarization-independent rib waveguides operating in telecommunication wavelengths with smaller dimensions varying the height, width, and etching depth using a mode solver program. Chan [25] also produced single-mode and polarization-independent equations for relatively small rib waveguides. Progress of research was not only restricted to silicon; it had been extended to polymers also like SU-8, which is an epoxy-based low-cost negative photoresist. Scientists from different groups like Beche et al. [26] designed and characterized single-mode rib waveguides made of SU-8 polymer and conveyed about the application of these waveguides in integrated optics with low optical losses; the work of

Pelletier et al. [27] dealt with SU-8 rib waveguides for sensing applications. The progress in experimentation was then extended to bent rib waveguides, which were used as fiber pigtails in coupled waveguides and in connecting different components that require lateral shift. Halir et al. [28] in 2007 analyzed bent single-mode rib waveguides (both rectangular and trapezoidal) along with their fabrication tolerances and proposed a procedure to calculate their minimum bending radius. Simultaneously, fabrication and characterization, i.e., to develop the designed rib structures using different methods came into play. Navalakhe et al. [29] were the first in India to fabricate and characterize their designed single-mode optical rib waveguide at wavelength 1.55 μm in SOI platform. In 2009, they fabricated and characterized straight and S-bend Si rib waveguide structures and showed experimentally that the bending radius of an asymmetrically etched S-bend waveguide can be ten times smaller than that of conventional symmetrically etched S-bend waveguides for similar optical losses [30]. For fabricating optical devices of extremely small dimensions, wire waveguide is an attractive as well as desirable platform. Vlasov et al. [31] fabricated single-mode SOI strip waveguides and bends and reported their measured propagation and bending losses for these waveguides. Fabrication by UV lithography process using chrome mask for SU-8 polymeric waveguides and their characterizations is done by Tung et al. [32]. In 2012, Tripathy et al. [33] investigated the sensitivity of SU-8 waveguides by introducing C- and S-shaped bends.

Along with the tremendous research on rib and wire waveguides and bends, researchers underwent design and fabrication of directional couplers composed of these waveguides and having wide use in the field of optoelectronic devices, various photonic devices like micro-ring resonators and power splitters. Quan et al. [34] in 2008 fabricated photonic wire waveguide-based directional coupler on SOI platform using electron beam lithography (EBL) and inductively coupled plasma (ICP) etching systems; simulation was done using finite-difference time-domain (FDTD) method, and they showed that fabricated results match with the simulated ones. Modeling of polarization-insensitive silicon wire-based directional coupler was presented by Pasaro et al. [35] in 2008 with a semi-analytical approach based on the coupled mode theory (CMT) and finite element method (FEM). In 2010, George et al. [36] studied on the design, fabrication, and characterization of directional couplers with symmetrically and asymmetrically etched S-bend silicon waveguides and reported that the device could be more compact if directional couplers with asymmetrically etched waveguide were used instead of conventional symmetrically etched bend waveguides without compromising much of the optical losses.

1.1.2 Micro-ring Resonator

Micro-ring resonator plays an important role in the development of modern integrated photonics. This device in its simplest form is an all-pass ring resonator or a notch filter configuration, where output of directional coupler is fed back to its input. The ring comprising the resonator is generally circular; however, in some cases, it may be racetrack configuration, in which the circular shape is elongated with a straight section along the coupling direction. In order to make the device compact, there is a need of small bend radius, which can be obtained if high-contrast waveguides (with strong optical confinement) were used. Niehusmann et al. [37] presented an ultrahigh quality factor circular ring resonator based on silicon wire waveguides and claimed that the obtained propagation loss of their fabricated device was the lowest at that time. Kiyat et al. [38] designed and fabricated racetrack resonator which is based on single-mode large cross-sectional rib waveguides and reported their measured results to be in a good compromise between good extinction ratios and high quality factors; the quality factor is the highest for resonators based on silicon-on-insulator rib waveguides during that time. Popovic et al. [39] were the first to demonstrate high-order micro-ring resonator add–drop filters meeting telecommunication specifications for dense wavelength division multiplexing (DWDM) applications that supported full free spectral range tunability. After that, ultra-compact fifth-order racetrack ring resonator optical filters based on submicron silicon photonic wires were demonstrated by Xia et al. [40] in 2007. In the same year, Huang et al. [41] experimentally realized rib-based ring resonator at telecommunication wavelength near 1550 nm. Racetrack resonator using SU-8 ridge waveguides was demonstrated by Dai et al. [42] for add–drop filters in 2008. Later, Prabhu et al. [43] conducted research for wavelength division multiplexing (WDM) on-chip interconnect applications; three-stage double-channel cascaded micro-rings integrated with grating couplers based on SOI rib waveguide were presented by Hu et al. [44] in 2010. Using parallel- and serial-coupled polymer ring resonators, Prajzler et al. [45] designed wavelength triplexer. Ring resonators can also be used as modulator and this had been demonstrated by Hu [46] in 2012. Recent works of Salleh et al. [47] showed the utility of SU-8-based ring resonators in bio-sensing applications. In 2013, Haldar et al. [48] proposed a theory on off-axis MRR in SOI platform, where they considered both single and multiple off-axis rings. Manzano et al. [49] discussed on the coupled sequence of micro-ring resonators while the resonance shift tuning according to the orientation of racetrack resonators was analyzed recently in 2018 by Castellan [50].

1.1.3 Photonic Crystal Structure

Photonic crystals or photonic bandgap materials have been one of the topics of interest during the last few decades. It is reported that effect of photons in case of photonic crystals is almost the same as semiconductors affect the properties of electrons. These are periodic structures with periods in the order of wavelength of light; the periodicity may be one-, two-, or three-dimensional. The potential applications of these devices include switching, in nonlinear optics, and as filters. They may be used in various forms like ring, multilayer stack or pillar slab. Functional devices like resonators, couplers, and splitters also utilize the concept of photonic crystal [51–57]. The first photonic crystal was proposed by Yablonovitch [58] in 1987; in 1991, he fabricated such crystals by drilling holes into a dielectric material of refractive index 3.6 [59]. Thereafter, a lot of studies on these devices were carried out with various numerical techniques and fabrication methods. Finite-difference time-domain method and finite element method were found to be widely used numerical techniques by which photonic crystals had been realized while for producing the nanometer-level crystals, electron beam lithography, x-ray lithography, and focused ion beam lithography were in the list. Although control of light in three-dimensional photonic crystals is in all directions in space; fabrication is not easy for these; one- and two-dimensional crystals are comparatively easy to fabricate and realize. Photonic crystal based on layer-by-layer fashion was reported by Ozbay et al. [60] for full photonic bandgap in 1996; Krauss et al. [61] developed two-dimensional structures at near-infrared wavelengths in the same year. Then there had been a keen interest among different research groups [62–64] to create optical microcavities; as a result, defect modes within photonic crystal came into play. It was Meade et al. [65] who gave the idea of using line defects in photonic crystals.

1.1.4 Power Splitter

Power splitters fall under the category of components which find their applications in optics communications, signal processing, and photonic integrated circuits (PICs) and, in general, are passive in nature. These are used for distributing and combining signals, so free choice of power splitting ratio plays an important role in the case of PIC applications. 3-dB splitters had gained attention from many researchers [66]. A low-loss cost-effective compact polarization- and wavelength-independent power splitter is the desire for every researcher, which is considered to be the ideal one. The design configuration of these components may be based on Y-branch, directional coupler, multi-mode interference, photonic crystal, subwavelength plasmonic waveguide, Mach–Zehnder, combination of Mach–Zehnder interferometer and multi-mode interference couplers, or ring resonator theories [67–75]. Splitters which are based on evanescent field coupling and multi-mode interference are strongly wavelength and polarization-dependent, mostly if the choice of material is of low-index contrast, thus cannot be used in broadband applications. Y-branch splitters though have less polarization and wavelength dependencies result in more excess loss due to the mismatch in mode field and wavefront in between the input and output branches; also, they have poor transmission at the bends and input and output ports. Photonic crystal-based splitters as optical interconnects have low transmission and bending loss with reduced crosstalk; however, high-resolution lithography and high aspect ratio etching are required for manufacturing where slight imperfection in feature sizes or surface conditions affect dispersion and scattering in photonic crystals. The configuration based on combined Mach–Zehnder interferometer and multi-mode interference has the advantage of being configurable but has polarization effects.

1.2 Motivation and Objectives

To conclude from review of literatures, it is found that most of the theoretical analyses of waveguides and waveguide-based devices are based on purely numerical techniques, such as beam propagation method (BPM) and finite-difference time-domain (FDTD) method. Although BPM or FDTD in their advanced forms are the most powerful computer simulation techniques to analyze devices with structural variations along the propagation direction and compute device parameters accurately, they are highly computation-intensive and require huge computer run-time and memory. On the other hand, the growing complexities of OICs demand a computationally faster method for determining the basic parameters of the waveguides and waveguide-related components, the building blocks of OIC, so that the overall simulation of the OIC (consisting of so many such components) does not become too complex and unmanageable in standard computers and work stations. Thus, there is a need to develop a simpler method other than BPM or FDTD to analyze waveguides and related components and compute their characteristics and necessary parameters for the overall system-level simulation with a higher speed and reasonable accuracy. A faster design method of integrated optics, therefore, needs the following features: (a) should be capable of computing the propagation constants for single-mode as well as multi-mode waveguides and also for coupled guides for both transverse electric (TE) and transverse magnetic (TM) polarizations of input light; (b) ability to determine the modal field profiles in single as well as coupled waveguides; (c) should be able to calculate the propagation losses in leaky waveguide structures such as waveguide bends; and (d) the method should be basically non-iterative in nature to make computation extremely fast.

For waveguide-based device fabrication, the fabrication methods, whether masked or maskless, have some pros and cons. Being maskless, direct-write lithography techniques like electron beam lithography, laser beam direct writing, and proton beam writing are capable of inexpensive rapid prototyping; however, these can never compete with the masked lithography techniques in terms of manufacturing throughput. Thus, no method can be said clearly superior; the choice of the method depends on the aspect of desired applications and fabrication tolerances. Also, the characterization setup should be chosen such that the cost is minimum without compromise in device yield.

The objectives of this monograph are the following:

(i) To design and analyze single-mode silicon and SU-8 waveguides using a method which is less computation-intensive, thus faster and occupy less computer memory than the existing commercially available softwares.
(ii) To fabricate and characterize the designed single-mode wire waveguides with cost-effective techniques and methods.
(iii) To develop on-chip photonic devices using the fabricated SU-8 wire waveguides

(a) Optical directional coupler
(b) Micro-ring resonator
(c) Photonic crystal structure on waveguide.

(iv) To develop a polarization-independent power splitter using single-mode silicon wire/rib waveguides for on-chip interconnect applications.

1.3 Outline of the Book

This monograph comprises of the design and analysis, fabrication and characterization of silicon, and SU-8 waveguides and waveguide-based devices. Chapter 2 deals with the theoretical studies on rib and wire waveguides made up of silicon and SU-8 polymer. Design of single-mode waveguides, bending loss computation of bent waveguides, lateral mode profile computation, design aspects of waveguide with slanted-etched wall are described in this portion. The experimental studies regarding the fabrication and characterization of SU-8 wire waveguides are presented in Chap. 3. Chapter 4 is on the design and development of SU-8 wire waveguide structures, viz. optical directional coupler, micro-ring resonator, and photonic crystal structure on waveguide. Fabrication and characterization of rib waveguide and the design and development of three-waveguide polarization-independent power splitter using these rib waveguides are elaborated in Chap. 5. Chapter 6 makes a summary of the work described so far and possible future scopes of the work.

References

1. B.C. Kress, *Field Guide to Digital Micro-optics*. eISBN: 9781628411843 (2014)
2. H. Kogelnik, Theory of optical waveguides, in *Guided-Wave Optoelectronics*, ed. by T. Tamir (Springer, Berlin, Heidelberg, 1988), pp. 7–88
3. S. Bhadra, A. Ghatak, *Guided Wave Optics and Photonic Devices* (CRC Press, Taylor & Francis Group, 2017)
4. G.P. Agrawal, *Lecture Slides, The Institute of Optics* (University of Rochester, 2008)
5. L. Vivien, *Recent Advances in Silicon Photonics*. https://indico.cern.ch/event/291295/. Accessed on Oct 2014
6. F. Grillot, Propagation loss in single-mode ultrasmall square silicon-on-insulator optical waveguides. J. Lightwave Technol. **24**, 891–896 (2006)
7. O. Kononchuk, B.Y. Nguyen, *Silicon-on-Insulator (SOI) Technology: Manufacture and Applications* (Elsevier, Amsterdam, 2014)
8. B. Yang, L. Yang, R. Hu, Z. Sheng, D. Dai, Fabrication and characterization of small optical ridge waveguides based on SU-8 polymer. J. Lightwave Technol. **27**, 4091–4096 (2009)
9. M. Nordstrom, D.A. Zauner, A. Boisen, J. Hubner, Single-mode waveguides with SU-8 polymer core and cladding for MOEMS applications. J. Lightwave Technol. **25**, 1284–1289 (2007)
10. P.K. Dey, P. Ganguly, A technical report on fabrication of SU-8 optical waveguides. J. Optics **43**, 79–83 (2014)
11. V.C. Pinto, P.J. Sousa, V.F. Cardoso, G. Minas, Optimized SU-8 processing for low-cost microstructures fabrication without cleanroom facilities. Micromachines **5**, 738–755 (2014)

12. *SU-8: A Versatile Material for MEMS Manufacturing* (2006). Available: http://gersteltec.ch/userfiles/1197911855.pdf
13. D. Dai, B. Yang, L. Yang, Z. Sheng, Design and fabrication of SU-8 polymer-based micro-racetrack resonators, in *Proceedings of the SPIE—The International Society for Optical Engineering*, vol. 7134 (2008), p. 713414
14. B.Y. Shew, C.H. Kuo, Y.C. Huang, Y.H. Tsai, UV-LIGA interferometer biosensor based on the SU-8 optical waveguide. Sens. Actuators A **120**, 383–389 (2005)
15. C.J. Robin, A. Vishnoi, K.N. Jonnalagadda, Mechanical behavior and anisotropy of spin-coated SU-8 thin films for MEMS. J. Microelectromech. Systems **23**, 168–180 (2014)
16. M. Joshi, N. Kale, R. Lal, V.R. Rao, S. Mukherji, A novel dry method for surface modification of SU-8 for immobilization of biomolecules in Bio-MEMS. Biosens. Bioelectron. **22**, 2429–243 (2007)
17. K. Gut, T. Herzog, Analysis and investigations of differential interferometer based on a polymer optical bimodal waveguide. Photonics Lett. Poland **7**, 56–58 (2015)
18. T. Herzog, K. Gut, Near field light intensity distribution analysis in bimodal polymer waveguide, in *Optical Fibers and Their Applications* (Lublin and Naleczow, Poland, 98160K, 2015)
19. A.L. Bogdanov, Use of SU-8 negative photoresist for optical mask manufacturing, in *Proceedings of the SPIE, Advanced Resist Technology Processing XVII*, vol. 3999 (2000), pp. 1215–1225
20. R.A. Soref, J. Schmidtchen, K. Petermann, Large single-mode rib waveguides in GeSi-Si and Si-on-SiO$_2$. IEEE J. Sel. Top. Quantum Electron. **27**, 1971–1974 (1991)
21. K. Petermann, Properties of optical rib-guides with large cross-section. Uberetragungstechnik Electron. Commun. **30**, 139–140 (1976)
22. S.P. Pogossian, L. Vescan, A. Vonsovici, The single-mode condition for semiconductor rib waveguides with large cross section. J. Lightwave Technol. **16**, 1851–1853 (1998)
23. O. Powell, Single-mode condition for silicon rib waveguides. J. Lightwave Technol. **20**, 1851–1855 (2002)
24. L. Vivien, S. Laval, B. Dumont, S. Lardenois, A. Koster, E. Cassan, Polarization-independent single-mode rib waveguides on silicon-on-insulator for telecommunication wavelengths. Opt. Commun. **210**, 43–49 (2002)
25. S.P. Chan, C.E. Png, S.T. Lim, G.T. Reed, V.M.N. Pasaro, Single-mode and polarization-independent silicon-on-insulator waveguides with small cross section. J. Lightwave Technol. **23**, 2103–2111 (2005)
26. B. Beche, N. Pelletier, E. Gaviot, J. Zyss, Single-mode TE$_{00}$–TM$_{00}$ optical waveguides on SU-8 polymer. Optics Commun. **230**, 91–94 (2004)
27. N. Pelletier, B. Beche, E. Gaviot, J. Zyss, Single-mode rib optical waveguides on SOG/SU-8 polymer and integrated Mach-Zehnder for designing thermal sensors. J. IEEE Sens. **6**, 565–570 (2006)
28. R. Halir, A. Ortega-Monux, J.G. Wangüemert-Perez, I. Molina-Fernández, P. Cheben, Fabrication tolerance analysis of bent single-mode rib waveguides on SOI. Opt. Quant. Electron. **38**, 921–932 (2007)
29. R.K. Navalakhe, N. Dasgupta, B.K. Das, Fabrication and characterizations of single mode optical waveguide, in *Silicon-On-Insulator, Photonics-2008: International Conference on Fiber Optics and Photonics*, IIT Delhi, India (2008), pp. 13–17
30. R.K. Navalakhe, N. Dasgupta, B.K. Das, Fabrication and characterization of straight and compact S-bend optical waveguides on a silicon-on-insulator platform. Appl. Opt. **48**, G125–G130 (2009)
31. Y.A. Vlasov, S.J. McNab, Losses in single-mode silicon-on-insulator strip waveguides and bends. Opt. Express **12**, 1622–1631 (2004)
32. K.K. Tung, W.H. Wong, E.Y.B. Pun, Polymeric optical waveguides using direct ultraviolet photolithography process. Appl. Phys. A Mater. Sci. Process. **80**, 621–626 (2005)
33. R. Tripathi, A. Prabhakar, S. Mukherji, Comparison of micro fabricated C and S bend shape SU-8 polymer waveguide of different bending diameters for maximum sensitivity, in *International Symposium on Physical and Technology Sensors*, Pune, India (2012), pp. 228–231

34. Y. Quan, P.D. Han, Q.J. Ran, F.P. Zeng, L.P. Gao, C.H. Zhao, A photonic wire-based directional coupler based on SOI. Opt. Commun. **281**, 3105–3110 (2008)
35. V.M.N. Pasaro, F.D. Olio, B. Timotijevic, G.Z. Mashanovich, G.T. Reed, Polarization-Insensitive directional couplers based on SOI. Wire Waveguides **2**, 6–9 (2008)
36. J.P. George, N. Dasgupta, B.K. Das, Compact integrated optical directional coupler with large cross section silicon waveguides, silicon photonics and photonic integrated circuits II, n *Proceedings of the SPIE Photonics Europe*, vol. 7719 (2010), p. 77191X
37. J. Niehusmann, A. Vorckel, P.H. Bolivar, T. Wahlbrink, W. Henschel, H. Kurz, Ultrahigh-quality-factor silicon-on-insulator microring resonator. Opt. Lett. **29**, 2861–2863 (2004)
38. I. Kiyat, A. Aydinli, N. Dagli, High-Q silicon-on-insulator optical rib waveguide racetrack resonators. Opt. Exp. **13**, 1900–1905 (2005)
39. M.A. Popovic, T. Barwicz, M.R. Watts, P.T. Rakich, L. Socci, E.P. Ippen, F.X. Kartner, H.I. Smith, Multistage high-order microring-resonator add–drop filters. Opt. Lett. **31**, 2571–2573 (2006)
40. F. Xia, M. Rooks, L. Sekaric, Y. Vlasov, Ultra-compact high order ring resonator filters using submicron silicon photonic wires for on-chip optical interconnects. Opt. Exp. **15**, 11934–11941 (2007)
41. Q. Huang, J. Yu, S. Chen, X. Xu, W. Han, Z. Fan, High Q microring resonator in silicon-on-insulator rib waveguides. Proc. SPIE Optoelectronic Devices and Integration II **6838**, 68380J (2007)
42. D. Dai, B. Yang, L. Yang, Z. Sheng, Design and fabrication of SU-8 polymer-based micro-racetrack resonators, in *Passive Components and Fiber-based Devices V, Proceedings of the SPIE, Asia-Pacific Optical Communications*, vol. 7134 (2008), p. 713414
43. A.M. Prabhu, A. Tsay, Z. Han, V. Van, Ultracompact SOI microring add–drop filter with wide bandwidth and wide FSR. IEEE Photon Technol Lett. **21**, 651–653 (2009)
44. Y. Hu, X. Xiao, Y. Zhu, H. Xu, L. Zhou, Y. Li, Z. Fan, Z. Li, Y. Yu, J. Yu, design, fabrication and characterization of cascaded SOI rib waveguide microring resonators, in *7th IEEE International Conference on Group IV Photonics* (2010), pp. 216–218
45. V. Prajzler, E. Strilek, J. Spirkova, V. Jerabek, Design of the novel wavelength triplexer using multiple polymer microring resonators. Radioengineering **21**, 258–263 (2012)
46. Y. Hu, X. Xiao, H. Xu, X. Li, K. Xiong, Z. Li, T. Chu, Y. Yu, J. Yu, High-speed silicon modulator based on cascaded microring resonators. Opt. Exp. **20**, 15079–15085 (2012)
47. M.H.M. Salleh, A. Glidle, M. Sorel, J. Reboud, J.M. Cooper, Polymer dual ring resonators for label-free optical biosensing using microfluidics. Chem. Commun. **49**, 3095–3097 (2013)
48. R. Haldar, S. Das, S.K. Varshney, Theory and design of off-axis microring resonators for high-density on-chip photonic applications. J. Lightwave Technol. **31**, 3976–3986 (2013)
49. F.R. Manzano, S. Biasi, M. Bernard, M. Mancinelli, T. Chalyan, F. Turri, M. Ghulinyan, M. Borghi, A. Samusenko, D. Gandol, R. Guider, A. Trenti, P.E. Larre, L. Pasquardini, N. Prltjaga, S. Mana, I. Carusotto, G. Pucker, L. Pavesi, Microring resonators and silicon photonics. MRS Adv. **1**, 3281–3293 (2016)
50. C. Castellan, A. Chalyan, M. Mancinelli, P. Guilleme, M. Borghi, F. Bosia, N.M. Pugno, M. Bernard, M. Ghulinyan, G. Pucker, L. Pavesi, Tuning the strain-induced resonance shift in silicon racetrack resonators by their orientation. Opt. Express **26**, 4204–4218 (2018)
51. Z. Ma, K. Ogusu, Channel drop filters using photonic crystal Fabry-Perot resonators. Opt. Commun. **284**, 1192–1196 (2011)
52. N. Yamamoto, T. Ogawa, K. Komori, Photonic crystal directional coupler switch with small switching length and wide bandwidth. Opt. Exp. **14**, 1223–1229 (2006)
53. J.B. Abad, A. Rodriguez, P. Bermel, S.G. Johnson, J.D. Joannopoulos, M. Soljacic, Enhanced nonlinear optics in photonic-crystal microcavities. Opt. Exp. **15**, 16161–16176 (2007)
54. T.B. Yu, M.H. Wang, X.Q. Jiang, Q.H. Liao, J.Y. Yang, Ultracompact and wideband power splitter based on triplephotonic crystal waveguides directional coupler. J. Opt. A **9**, 37–42 (2007)
55. R.K. Ramakrishnan, S. Warrier, P. Angadikkunnath, A. Shenoy, S. Talabatulla, Analysis of nonlinear optical properties of photonic crystal beam splitters, in *Proceedings of the SPIE Photonics Europe*, vol. 7713 (2010), p. 77131X

56. V.D. Kumar, T. Srinivas, A. Selvarajan, Investigation of ring resonators in photonic crystal circuits. Photonics Nanostruct. Fundam. Appl. **2**, 199–206 (2004)
57. T. Sreenivasulua, V. Rao, T. Badrinarayana, G. Hegde, T. Srinivas, Photonic crystal ring resonator based force sensor: design and analysis. Optik **155**, 111–120 (2018)
58. E. Yablonovitch, Inhibited spontaneous emission in solid-state physics and electronics. Phys. Rev. Lett. **58**, 2059–2062 (1987)
59. E. Yablonovitch, T.J. Gmitter, Photonic band structure: the face-centered-cubic case employing nonspherical atoms. Phys. Rev. Lett. **67**, 2295–2298 (1991)
60. E. Ozbay, Layer-by-layer photonic crystals from microwave to far-infrared frequencies. J. Opt. Soc. Am. B **13**, 1945–1955 (1996)
61. T.F. Krauss, R.M.D.L. Rue, S. Brand, Two-dimensional photonic-bandgap structures operating at near infrared wavelengths. Nature **383**, 699–702 (1996)
62. O. Painter, J. Vuckovic, A. Scherer, Defect modes of a two-dimensional photonic crystal in an optically thin dielectric slab. J. Opt. Soc. Am. B **16**, 275–285 (1999)
63. D.J. Rippin, K. Lim, G.S. Petrich, P.R. Villeneuve, S. Fan, E.R. Thoen, J.D. Joannopoulos, E.P. Ippen, L.A. Kolodziejski, One-dimensional photonic bandgap microcavities for strong optical confinement in GaAs and GaAs/Al O semiconductor waveguides. J. Lightwave Technol. **17**, 2152–2160 (1999)
64. M. Tokushima, H. Kosaka, A. Tomita, H. Yamada, Lightwave propagation through a 120° sharply bent single-line-defect photonic crystal waveguide. Appl. Phys. Lett. **76**, 952–954 (2000)
65. R.D. Meade, A. Devenyi, J.D. Joannopoulos, O.L. Alerhand, D.A. Smith, K. Kash, Novel applications of photonic band gap materials: low-loss bends and high Q cavities. J. Appl. Phys. **75**, 4753–4755 (1994)
66. Z. Li, J. Xing, B. Yang, Y. Yu, Broadband optical beam power splitter for wavelength dependent light circuits on silicon substrates, in *International Conference on Optoelectronics and Microelectronics*, Harbin (2013), pp. 177–179
67. K.K. Chung, H.P. Chan, P.L. Chu, A 1 × 4 polarization and wavelength independent optical power splitter based on a novel wide-angle low-loss Y-junction. Opt. Commun. **267**, 367–372 (2006)
68. I. Park, H.S. Lee, H.J. Kim, K.M. Moon, S.G. Lee, B.H. O, S.G. Park, E.H. Lee., Photonic crystal power-splitter based on directional coupling. Opt. Exp. **12**, 3599–3604 (2004)
69. Y. Sakamaki, T. Saida, T. Hashimoto, H. Takahashi, Low-loss Y-branch waveguides designed by wavefront matching method. J. Lightwave Technol. **27**, 1128–1134 (2009)
70. Y. Zhang, L. Liu, X. Wu, L. Xu, Splitting-on-demand optical power splitters using multimode interference (MMI) waveguide with programmed modulations. Opt. Commun. **281**, 426–432 (2008)
71. K.B. Chung, J.S. Yoon, Properties of a 1 × 4 optical power splitter made of photonic crystal waveguides. Opt. Quantum Electron. **35**, 959–966 (2003)
72. J. Wang, X. Guan, Y. He, Y. Shi, Z. Wang, S. He, P. Holmstrom, L. Wosinski, L. Thylen, D. Dai, Sub-μm^2 power splitters by using silicon hybrid plasmonic waveguides. Opt. Exp. **19**, 838–847 (2011)
73. A. Ghaffari, F. Monifi, M. Djavid, M.S. Abrishamian, Photonic crystal bends and power splitters based on ring resonators. Opt. Commun. **281**, 5929–5934 (2008)
74. L.W. Cahill, The modelling of integrated optical power splitters and switches based on generalised Mach-Zehnder devices. Opt. Quantum Electron. **36**, 165–173 (2004)
75. V. Prajzler, H. Tuma, J. Spirkova, V. Jerabek, Design and modeling of symmetric three branch polymer planar optical power dividers. Radioengineering **22**, 233–239 (2013)

Chapter 2
Theoretical Studies on Silicon and SU-8 Waveguides

2.1 Introduction

In this chapter, theoretical studies regarding the design and analysis of silicon and SU-8 polymer waveguides are presented. The study starts with the design of single-mode optical wire waveguide at a transmitting wavelength of 1.55 μm. Effective index-based matrix method (EIMM) was used for this purpose, which is a two-step process. In the first step, effective index method (EIM) was used for vertical refractive index profile of waveguide, and then for the resulted lateral index profile, a transfer matrix method (TMM) was applied. The lateral mode profiles of wire for both transverse electric (TE) and transverse magnetic (TM) polarizations were also calculated using this approach. Bending losses of bend wire was computed using transfer matrix method along with conformal mapping technique. The analysis was then extended to silicon wire waveguide with slanted-etched wall and design of large cross-section silicon rib waveguide. The computed results of this semi-analytical EIMM were validated with commercially available Optiwave 2D-FDTD method. Although this EIMM method had already been tested for Titanium-indiffused lithium niobate (Ti:LiNbO$_3$) waveguides and LiNbO$_3$ photonic wires [1–5], it had not been applied previously, in its present form, for silicon or polymer (SU-8) wire and rib waveguides.

2.2 Design of Single-Mode Wire Waveguide

The work compiled in this section describes the design of single-mode wire waveguide at 1.55 μm transmitting wavelength using EIMM. Figure 2.1 shows the schematic of the wire waveguide where t is the oxide thickness, h and w are the respective thickness and width of the wire; n_1, n_b, and n_c are the refractive indices of the core (Si/SU-8), oxide (SiO$_2$) and cover (air), respectively. As mentioned, analysis of this waveguide was accomplished by two steps: for the vertical refractive index

© The Author(s), under exclusive license to Springer Nature Singapore Pte Ltd. 2020
S. Samanta et al., *Photonic Waveguide Components on Silicon Substrate*,
SpringerBriefs in Applied Sciences and Technology,
https://doi.org/10.1007/978-981-15-1311-4_2

Fig. 2.1 Schematic of silicon/SU-8 photonic wire (reprinted with permission from [10]. ©2015 Elsevier GmbH)

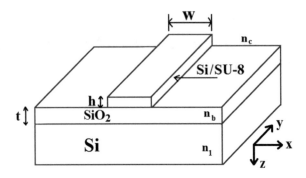

profile, effective index method [6] was used and then for the resulted lateral index profile, a transfer matrix method [7–9] was applied.

The scalar wave equation for the vertical modes of the waveguide can be written as:

$$\frac{\partial^2 \Psi}{\partial x^2} + \frac{\partial^2 \Psi}{\partial z^2} + \left[k^2 n^2(x, z) - \beta^2\right]\Psi = 0 \tag{2.1}$$

where Ψ is the field amplitude, and $n(x, z)$ is the refractive index of the waveguide. Applying appropriate boundary conditions at each interface of the waveguide structure in the depth direction of the waveguide (i.e., Z and $\partial X/\partial x$ are continuous at each interfaces for transverse electric (TE) mode, and Z and $(1/n^2)\partial X/\partial x$ are continuous at each interfaces for transverse magnetic (TM) mode; where $\psi = X(x)Z(z)$; $X(x)$ and $Z(z)$ being the electric/magnetic field along x and z directions, respectively), the modal dispersion equation for guided modes can be obtained; the derivation is as follows:

For TE Mode,

The field components of waveguide modes for TE polarization are E_x, H_z, and H_y. So, from Eq. (2.1), we get:

$$\frac{\partial^2 E_x}{\partial x^2} + \frac{\partial^2 E_x}{\partial z^2} + \left[\kappa^2 n^2(x, z) - \beta^2\right]E_x = 0; \quad \begin{array}{l}\text{where, } k = 2\pi/\lambda; \\ \lambda \text{ is the wavelength of light}\end{array}$$

Again, by assuming separation of variables:

$$E_x(x, z) = X(x)\, Z(z)$$

$$\text{Therefore,} \quad Z\frac{\partial^2 X}{\partial x^2} + X\frac{\partial^2 Z}{\partial z^2} + \left[\kappa^2 n^2(x, z) - \beta^2\right]X_z = 0$$

$$\text{or,} \quad \frac{1}{X}\frac{\mathrm{d}^2 X}{\mathrm{d}x^2} + \frac{1}{Z}\frac{\mathrm{d}^2 Z}{\mathrm{d}z^2} + \left[\kappa^2 n^2(x, z) - \beta^2\right] = 0$$

Fig. 2.2 Waveguide
cross-section

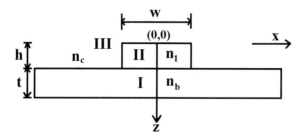

Now, we can write:

$$\frac{1}{X}\frac{d^2 X}{dx^2} + \left[\kappa^2 n_{eff,m}^2(x) - \beta^2\right] = 0 \tag{2.2}$$

$$\frac{1}{Z}\frac{d^2 Z}{dz^2} + \kappa^2\left[n^2(x, z) - n_{eff,m}^2(x)\right] = 0 \tag{2.3}$$

where $n_{eff,m}(x)$ varies only with respect to x; z dependency has been neglected. Since we are interested to apply effective index method in z-direction, Eq. (2.3) is our main concern.

Now, the waveguide cross-section is divided into three regions as shown in Fig. 2.2. Considering region I, and from Eq. (2.3), we get:

$$\frac{1}{Z}\frac{d^2 Z}{dz^2} + \kappa^2\left[n_b^2 - n_{eff,m}^2\right] = 0$$

$$\text{or,} \quad \frac{1}{Z}\frac{d^2 Z}{dz^2} + \kappa^2\left[n_{eff,m}^2 - n_b^2\right] = \gamma_1^2 \tag{2.4}$$

where $\gamma_1^2 = \kappa^2\left[n_{eff,m}^2 - n_b^2\right]$; and n_b is the substrate refractive index.

Now, the solution of Eq. (2.4) can be written as:

$$Z = c_1 e^{\gamma_1 z} + c_2 e^{-\gamma_1 z} \tag{2.5}$$

where c_1 and c_2 are constants.

Since the substrate thickness t is infinitely large, we assume no wave propagation in this region (as $z \to \infty$, $Z \to 0$). Thus, $c_1 = 0$, and Eq. (2.5) yields:

$$Z = c_2 e^{-\gamma_1 z} \tag{2.6}$$

In region II, $\frac{1}{Z}\frac{d^2 Z}{dz^2} + \kappa^2\left[n_1^2 - n_{eff,m}^2\right] = 0$

$$\text{or,} \quad \frac{d^2 Z}{dz^2} = -\gamma_2^2 Z \tag{2.7}$$

where $\gamma_2^2 = \kappa^2\left[n_1^2 - n_{\text{eff},m}^2\right]$.

The solution of Eq. (2.7) may be written as:

$$Z = c_3 \cos(\gamma_2 z) + c_4 \sin(\gamma_2 z)$$

where c_3 and c_4 are nonzero constants, since waveguide modes are propagating in this region.

Now, applying boundary conditions: at $z = h$, Z and $\frac{\partial Z}{\partial z}$ are continuous, we obtain:

$$c_2 e^{-\gamma_1 h} = c_3 \cos(\gamma_2 h) + c_4 \sin(\gamma_2 h) \tag{2.8}$$

and

$$-\gamma_1 c_2 e^{-\gamma_1 h} = -\gamma_2 c_3 \sin(\gamma_2 h) + \gamma_2 c_4 \cos(\gamma_2 h) \tag{2.9}$$

Dividing (2.9) by (2.8) gives:

$$\frac{c_3}{c_4} = -\frac{\gamma_2 \cos(\gamma_2 h) + \gamma_1 \sin(\gamma_2 h)}{\gamma_1 \cos(\gamma_2 h) - \gamma_2 \sin(\gamma_2 h)} \tag{2.10}$$

In region III, $\frac{1}{Z}\frac{d^2 Z}{dz^2} + \kappa^2\left[n_c^2 - n_{\text{eff},m}^2\right] = 0$

$$\text{or,} \quad \frac{d^2 Z}{dz^2} = \gamma_3^2 Z \tag{2.11}$$

where $\gamma_3^2 = \kappa^2\left[n_{\text{eff},m}^2 - n_c^2\right]$; n_c is the refractive index of the substrate.

Now, the solution of Eq. (2.11) can be written as:

$$Z = c_5 e^{\gamma_3 z} + c_6 e^{-\gamma_3 z}$$

Since superstrate thickness is infinitely large and we assume no wave propagation in this region (as $z \to$, $Z \to 0$). Thus, $c_6 = 0$, and Eq. (2.11) yields:

$$Z = c_5 e^{-\gamma_3 z} \tag{2.12}$$

Applying boundary conditions: at $z = 0$, Z and $\frac{\partial Z}{\partial z}$ are continuous, we obtain:

$$c_5 = c_3 \tag{2.13}$$

and

$$\gamma_3 c_5 = \gamma_2 c_4 \tag{2.14}$$

From (2.13) and (2.14),

$$\frac{c_3}{c_4} = \frac{\gamma_2}{\gamma_3} \tag{2.15}$$

Solving Eqs. (2.10) and (2.15), one may obtain :

$$\kappa h \sqrt{n_1^2 - n_{\text{eff},m}^2} = m\pi + \sum_{x=b,c} \tan^{-1}\left[\frac{\sqrt{n_{\text{eff},m}^2 - n_x^2}}{\sqrt{n_1^2 - n_{\text{eff},m}^2}}\right] \tag{2.16}$$

where $m = 0, 1, 2, 3, \ldots$.

For TM mode,

Z and $(1/n^2)\partial Z/\partial z$ are continuous at each interfaces.

Applying the boundary conditions and by the same approach as applied for TE mode, we obtain:

$$\kappa h \sqrt{n_1^2 - n_{\text{eff},m}^2} = m\pi + \sum_{x=b,c} \tan^{-1}\left[\left(\frac{n_1}{n_x}\right)^2 \sqrt{\frac{n_{\text{eff},m}^2 - n_x^2}{n_1^2 - n_{\text{eff},m}^2}}\right] \tag{2.17}$$

Thus, in general,

$$\kappa h \sqrt{n_1^2 - n_{\text{eff},m}^2} = m\pi + \sum_{x=b,c} \tan^{-1}\left[\left(\frac{n_1}{n_x}\right)^{2\rho} \sqrt{\frac{n_{\text{eff},m}^2 - n_x^2}{n_1^2 - n_{\text{eff},m}^2}}\right] \tag{2.18}$$

Now, effective refractive indices corresponding to different vertically guided modes can be found by solving this dispersion Eq. (2.18) numerically. It transfers the 2-D refractive index profile of the wire waveguide to an equivalent 1-D lateral effective index profile. Here, $\rho = 0$ for the TE mode while it is 1 for TM mode; $n_{\text{eff},m}$ is the effective index of the guided mode with mode number m (=0, 1, 2, ...); k (=$2\pi/\lambda$) is the free space propagation constant. Thus, after solving Eq. (2.18), one can determine the effective index of photonic wire for a fixed thickness (h) corresponding to any vertically guiding mode; or in other words, single-mode confinement in vertical direction of the waveguide can be ensured by controlling the thickness of waveguide core.

In the second step of the analysis, the lateral step effective refractive index profile was used for transfer matrix method. For this, we considered the wire waveguide as a layered structure [7] as shown in Fig. 2.3.

For TE mode, the field components are E_x, H_z and H_y. For a number (N) of step refractive index layers (each of thickness d_i and refractive index n_i) and for an incident plane wave in the first layer, the electric field associated with each layer may be represented in the following form:

Fig. 2.3 Waveguide as layered structure

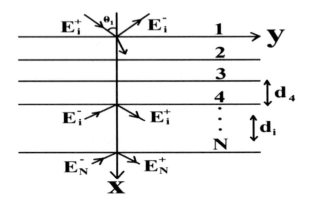

$$E_i = e_i^+ E_i^+ e^{i\Delta_i} e^{i(\omega t - \kappa_i \cos\theta_i x - \beta y)} + e_i^- E_i^- e^{-i\Delta_i} e^{i(\omega t + \kappa_i \cos\theta_i x - \beta y)} \tag{2.19}$$

At $y = 0, t = 0$; $\Delta_1 = \Delta_2 = 0$; $\Delta_3 = k_3 d_2 \cos\theta_3$;

$\Delta_i = k_i \cos\theta_i (d_2 + d_3 + \cdots + d_{i-1})$;

$k_i = \kappa_0 n_i = \dfrac{\omega}{c} n_i$; $\beta_i = k_i \sin\theta_i$; $i = 1, 2, \ldots, N$

E_i^+ and E_i^- are the electric field amplitudes of transmitted and reflected propagating waves, respectively; e_i^+ and e_i^- are the unit vectors along with the electric field directions; β being the propagation constant is an invariant of the structure. The boundary conditions at the interface of i and $(i + 1)$, the layers are as follows:

$$E_i^+ + E_i^- = E_{i+1}^+ + E_{i+1}^- \tag{2.20}$$

$$\left.\frac{\partial E_i}{\partial x}\right|_{x=d_i} = \left.\frac{\partial E_{i+1}}{\partial x}\right|_{x=d_{i+1}} \tag{2.21}$$

From Eq. (2.21), we obtain:

$$E_i^+ e^{i\Delta_i} - E_i^- e^{-i\Delta_i} = \frac{\kappa_{i+1}\cos\theta_{i+1}}{\kappa_i\cos\theta_i}[E_{i+1}^+ e^{i\Delta_{i+1}} e^{-j\kappa_{i+1}\cos\theta_{i+1}(d_1+\cdots d_i)}$$
$$- E_{i+1}^- e^{-i\Delta_{i+1}} e^{j\kappa_{i+1}\cos\theta_{i+1}(d_1+\cdots d_i)}] \tag{2.22}$$

Solving Eqs. (2.20) and (2.22), we get two equations as follows:

$$E_i^+ = \frac{1}{2}\left(1 + \frac{n_{i+1}\cos\theta_{i+1}}{n_i\cos\theta_i}\right)e^{i\Delta_{i+1}} e^{i[\partial_i - \kappa_{i+1}(d_1+\cdots d_i)\cos\theta_{i+1}]} E_{i+1}^+$$
$$+ \frac{1}{2}\left(1 - \frac{n_{i+1}\cos\theta_{i+1}}{n_i\cos\theta_i}\right)e^{-i\Delta_{i+1}} e^{i[\partial_i + \kappa_{i+1}(d_1+\cdots d_i)\cos\theta_{i+1}]} E_{i+1}^- \tag{2.23}$$

and

$$
\begin{aligned}
E_i^- = {} & \frac{1}{2}\left(1 - \frac{n_{i+1}\cos\theta_{i+1}}{n_i\cos\theta_i}\right)e^{i\Delta_{i+1}}e^{-i[\partial_i + \kappa_{i+1}(d_1+\cdots d_i)\cos\theta_{i+1}]}E_{i+1}^+ \\
& + \frac{1}{2}\left(1 + \frac{n_{i+1}\cos\theta_{i+1}}{n_i\cos\theta_i}\right)e^{-i\Delta_{i+1}}e^{-i[\partial_i - \kappa_{i+1}(d_1+\cdots d_i)\cos\theta_{i+1}]}E_{i+1}^-
\end{aligned} \tag{2.24}
$$

where $\partial_i = k_i d_i \cos\theta_i$.

Now, (2.23) and (2.24) yield the following matrix relation:

$$
\begin{pmatrix} E_1^+ \\ E_1^- \end{pmatrix} = \frac{1}{t_i}\begin{pmatrix} e^{i\partial_i} & r_i e^{i\partial_i} \\ r_i e^{-i\partial_i} & e^{-i\delta_i} \end{pmatrix}\begin{pmatrix} E_{i+1}^+ \\ E_{i+1}^- \end{pmatrix}
$$

$$
\text{or,}\quad \begin{pmatrix} E_1^+ \\ E_1^- \end{pmatrix} = T_1\begin{pmatrix} E_2^+ \\ E_2^- \end{pmatrix} = \cdots = T_1 T_2 \ldots T_{N-1}\begin{pmatrix} E_N^+ \\ E_N^- \end{pmatrix};
$$

$$
\text{where,}\quad T_i = \frac{1}{t_1}\begin{pmatrix} e^{i\partial_i} & r_i e^{i\partial_i} \\ r_i e^{-i\partial_i} & e^{-i\partial_i} \end{pmatrix} \tag{2.25}
$$

r_i and t_i represent the Fresnel amplitude reflection and transmission coefficients, respectively, at the ith interface and are given as:

$$
r_i = (n_i\cos\theta_i - n_{i+1}\cos\theta_{i+1})/(n_i\cos\theta_i + n_{i+1}\cos\theta_{i+1})
$$

$$
t_i = (2n_i\cos\theta_i)/(n_i\cos\theta_i + n_{i+1}\cos\theta_{i+1})
$$

For TM mode, the field components are H_x, E_z and E_y. Thus, for a number (N) of step refractive index layers (each of thickness d_i and refractive index n_i) and for an incident plane wave in the first layer, the magnetic field associated with each layer may be represented in the following form:

$$
H_i = \frac{n}{\mu_0 c}\left[e_i^+ H_i^+ e^{i\Delta_i} e^{i(\omega t - \kappa_i \cos\theta_i x - \beta y)} + e_i^- H_i^- e^{-i\Delta_i} e^{i(\omega t + \kappa_i \cos\theta_i x - \beta y)}\right] \tag{2.26}
$$

where H_i^+ and H_i^- are the magnetic field amplitudes of transmitted and reflected propagating waves, respectively. Now, the boundary conditions at the interface of i and $(i+1)$ layers are:

$$
H_i^+ + H_i^- = H_{i+1}^+ + H_{i+1}^-; \quad \frac{1}{n_i^2}\frac{\partial H_i}{\partial x} = \frac{1}{n_{i+1}^2}\frac{\partial H_{i+1}}{\partial x} \tag{2.27}
$$

By solving Eqs. (2.26) and (2.27), we obtain the same equation as (2.25) with respective Fresnel amplitude reflection and transmission coefficients as follows:

$$
r_i = (n_{i+1}\cos\theta_i - n_i\cos\theta_{i+1})/(n_{i+1}\cos\theta_i + n_i\cos\theta_{i+1})
$$

$$
t_i = (2n_i\cos\theta_i)/(n_{i+1}\cos\theta_i + n_i\cos\theta_{i+1})
$$

The above method together with prism-coupling concept may be applied to the 1-D effective refractive index profile of the wire waveguide to obtain the propagation constant of the guided modes. A higher index layer (n_0) was considered as first layer and excitation efficiency ($|E_g^+/E_1^+|^2$) of the guiding layer (highest refractive index layer, gth, of the structure) was computed by matrix method for different incident angles. Out of different incident angles, only a few will excite the guided modes, and in excitation efficiency versus propagation constant/incident angle plot, Lorentzian resonance peaks would appear. One may obtain propagation constants of the guided modes of the waveguide from the peak positions, and full-width-at-half-maxima of these peaks will indicate the radiation loss of the modes [8, 9]. Thus, single-mode wire waveguide can be designed by choosing the proper width of the wire, which results into only one sharp resonance peak.

2.3   Bending Loss Computation of Bent Wire

Curved optical waveguides are used to connect fiber pigtails with coupled waveguides in a directional coupler at the inputs and outputs. These are also to interconnect different integrated optic components on the substrate where lateral shifts are required. Si/SU-8 photonic wire waveguide bends are attractive in this context, since these waveguides can be bent with extremely small curvatures of less than a few micrometers of bending radius. When light propagates through these bent waveguides, the bending loss will occur. In this work, the bending loss of bent wire waveguide computation was done using EIMM along with a conformal mapping technique. First of all, conformal mapping technique was used to convert effective index profile of bent waveguide into an equivalent straight waveguide of asymmetric refractive index profile (n_{eq}); this was followed by transfer matrix method to determine the resonant peak of the propagation constant of the waveguide. Figure 2.4 shows the conformal mapping transformation of bent waveguide having refractive index n_{eff} (x, y) in (x, y) plane, which is converted to equivalent straight waveguide in (u, v) plane with

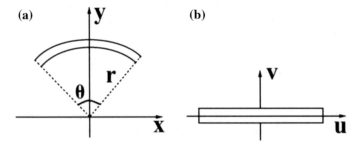

Fig. 2.4 Conformal mapping transformation from **a** bent waveguide to equivalent and **b** straight waveguide

modified equivalent refractive index $n_{eq}(u, v)$. Here, the optical distance across this bent waveguide $r\theta$ is larger at the outer edge compared to the inner one; r being the radii of curvature of the bent waveguide central line; θ being the angle of incidence.

Thus, the conformal transformation for a two-dimensional scalar wave equation for a uniformly curved waveguide is the mapping between two complex planes, i.e., from $(z = x + iy)$ plane to $(w = u + iv)$ plane. In other words, $w = f(x, y)$; f being an analytical function. Now, the required transformation for the purpose is [11]:

$$u = r \ln\left(1 + \frac{x}{r}\right) \quad \text{and} \quad n_{eq}(u) = \exp\left(\frac{u}{r}\right) n_{eff}(x) \tag{2.28}$$

In Eq. (2.28), x was measured from the center of the waveguide, $n_{eff}(x)$ being the effective refractive index profile. It may be pointed out that this transformation is not restricted to large bending radii, but is also applicable for smaller radii comparable to the width of the wire. As the optical distance across this bent waveguide $r\theta$ is larger at the outer edge compared to the inner one, for the equivalent straight waveguide, there will be some radiation loss where refractive index is higher. Next, transfer matrix method was applied on $n_{eq}(u)$; Lorentzian-shaped resonance curve was obtained from the computation of coupling efficiency as a function of propagation constant (β). As full-width-at-half-maximum (Γ) of the resonance peak is twice the imaginary part of propagation constant, bending loss (BL) of the bent waveguide can be computed by the following formula:

$$BL \text{ (in dB/unit length)} = 4.34 \, \Gamma \tag{2.29}$$

2.4 Lateral Mode Profile Computation of Photonic Wire

To obtain the lateral electric and magnetic field profiles of the guided modes of the waveguide, Eqs. (2.19) and (2.26) were used to compute E_i and H_i in each layer, respectively. The propagation constant β of the guided modes, determined by the method discussed in Sect. 2.2, was used for this computation. The method involved only multiplication of 2×2 matrices, hence extremely fast, although the matrices contained complex elements. $\sum_{i=1}^{N} E_i^2$ as a function of x would give us the mode profile of each guided mode of the waveguide.

2.5 Design Aspects of Wire Waveguide with Slanted Wall

The analysis of wire waveguides were extended for trapezoidal structure, as shown in Fig. 2.5, with a side angle of 35.26°. For all practical cases, to get perfectly vertical side wall of the wire waveguides is impossible; the wall will be slanted by some

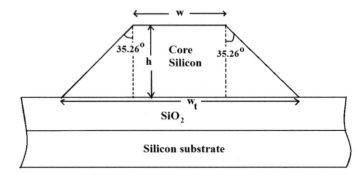

Fig. 2.5 Schematic of wire waveguide with slanted wall (reprinted with permission from [10]. ©2015 Elsevier GmbH)

angle. The angle chosen in Fig. 2.5 arises from anisotropic chemical etching with the waveguide aligned to the (110) crystal direction. Some researchers [12, 13] have taken average width of the trapezoidal structure to analyze the properties of waveguide. In this work, the slanted sidewalls of the waveguide were discretized into a number of rectangular grids, and for each grid (of different h), effective refractive indices were computed. The final outcome was a lateral refractive index profile which was graded at the two sides due to slanted side walls and step-index core (of constant h) was sandwiched in between. This lateral profile was then used in transfer matrix method to analyze the waveguide.

2.6   Design of Large Cross-Section Silicon Rib Waveguide

Single-mode condition for large cross-section Si rib waveguides were also determined by effective index-based matrix method. The rib waveguide structure in SOI platform is shown in Fig. 2.6. In the core region of the waveguide, the thickness of the top silicon layer was taken as H, whereas h_s represents the slab layer thickness which is always smaller than H. In the first step, effective index method was applied along

Fig. 2.6 Schematic of SOI
rib waveguide (reprinted
with permission from [10]
©2015 Elsevier GmbH)

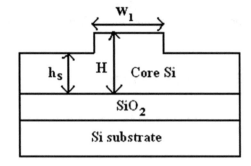

the depth direction of the waveguide. Slab region (of thickness h_s) produced a planar multi-mode (in depth direction) waveguide with effective index of the fundamental mode $n_{eff,slab}$. The core region of the waveguide (of thickness H) also supported a number of vertical modes with effective refractive index of the fundamental mode $n_{eff,0}$. One has to choose the proper combination of H and h so that $n_{eff,0}$ is always greater than $n_{eff,slab}$, and all the other higher-order modes in the core region is less than $n_{eff,slab}$. As a result, the higher-order vertical modes supported in the central region would couple into the fundamental slab mode of the rib-side regions. This lateral leakage ensures that these modes have high propagation losses, yielding an effectively single-mode optical waveguide in the vertical direction. Finally, matrix method was applied to determine the width of the rib waveguide for single-mode operation in the lateral direction. Effective refractive indices of the fundamental modes of slab and rib regions were used for the purpose.

2.7 Computed Results

All the computer programs written in this work were in Visual C++ and the refractive indices of Si, SU-8, SiO$_2$, and air were taken as 3.477, 1.574, 1.447, and 1.000, respectively, at 1.55 µm transmitting wavelength [14–17]. The dispersion equation (Eq. 2.18) was solved numerically for different modes of TE and TM polarizations to obtain the effective refractive indices. The computed results for Si and SU-8 photonic wires of first four modes are shown in Fig. 2.7. One may notice from these results that the Si waveguide supports only fundamental vertical mode ($m = 0$) within the film thickness 25.2–270.9 nm for TE mode and 106.1–353.2 nm for TM mode, while single-mode SU-8 waveguide operation is observed for thickness (h) lying in between $430 \le h$ (nm) ≤ 1720 and $560 \le h$ (nm) ≤ 1850, for TE and TM mode, respectively. To design single-mode Si wire waveguide at 1.55 µm wavelength of light, we chose top Si layer thickness (h) as 250 nm, which supported one TE mode and one TM mode. Transfer matrix method, as discussed in previous section was applied to determine the propagation constants for different widths (w) of the waveguide. The effective refractive indices for this film thickness were 2.925 for TE mode and 2.179 for TM mode (Fig. 2.7). In case of SU-8 waveguide, the chosen thickness was 1350 nm. In all calculations, the n_0 value was chosen equal to the maximum refractive index of the waveguide structure (n_{eff}) and gap d_1 (distance between the added layer and guided region) was increased until the limiting values of mode propagation constants were obtained. In the present work, optimum results were determined for $d_1 \sim 0.7$ µm.

The computed results of different widths of Si and SU-8 waveguides for TE propagation constants are shown in Fig. 2.8, while TM propagation constants are given in Fig. 2.9. We found that Si wire behaved as single-mode waveguide within widths 13–290 nm and 10–470 nm for TE and TM mode, respectively. SU-8 waveguide operated in single-mode condition between widths 60–670 nm for TE mode and

Fig. 2.7 Variation of
effective refractive indices
for different *h* for **a** Si wire
(reprinted with permission
from [10]. ©2015 Elsevier
GmbH). **b** SU-8 wire

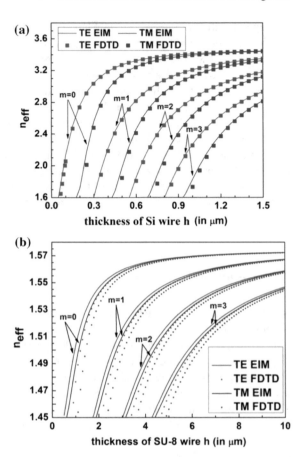

80–730 nm for TM mode. Thus, in our computations, we had chosen $w = 250$ nm and 600 nm to get single-mode (in lateral direction) silicon and SU-8 wire waveguide, respectively. Typical excitation efficiency versus propagation constant distribution of Si wire with a thickness and width of 250 nm for TE mode is shown in Fig. 2.10. The guided mode propagation constant of this TE fundamental mode is 9.5193 μm^{-1}.

It may be mentioned that during our computation the n_{eff} values for different *h* values (Fig. 2.7) were also compared with the results of 2D-FDTD simulator. A very good agreement of the results was obtained except near the mode cut-off, where n_{eff} was slightly higher when computed by EIM. This is obvious, since separation of variables does not hold well near cut-off. Physically, the mode is spread in the cladding, while separation of variables requires strong confinement. In our single-mode waveguide design process, we therefore avoided the '*h*' values very close to cut-off, which improved the overall accuracy of EIMM.

Fig. 2.8 Comparison of
computed propagation
constants for different widths
of waveguide by EIMM and
FDTD methods for TE mode
a Si waveguide (reprinted
with permission from [10].
©2015 Elsevier GmbH).
b SU-8 waveguide

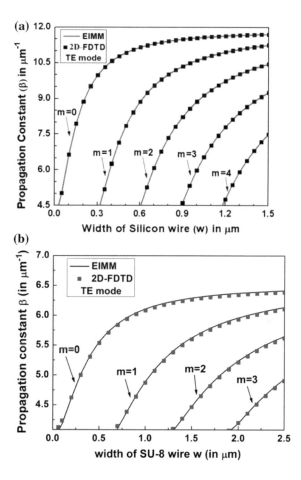

Since the propagation constant value of the guided mode was known, the lateral mode profile was computed from Eq. (2.19). Typical results for normalized TE and TM modes for Si wire waveguides are shown in Fig. 2.11. As the effective refractive indices for TE mode is greater than TM mode, TE mode profile is slightly more confined within the waveguide. A comparative result of the normalized mode profile between EIMM and FDTD for TE mode Si waveguide is shown in Fig. 2.12, which once again matches fairly well. Similar technique may also be used to obtain mode profile for other waveguide structures, such as bent or coupled waveguides.

It may be pointed out that for high-contrast Si (as well as SU-8) waveguide structures, rigorous full-vectorial BPM yields more accurate results than EIMM, although the former takes huge computation effort. Full-vectorial techniques also take into account mode hybridization (anticrossings). So, one may initially extract the design parameters of the waveguide-based devices with fair accuracy using EIMM, and then fine-tune those parameters using a full-vectorial technique.

Fig. 2.9 Comparison of computed propagation constants for different widths of waveguide by EIMM and FDTD methods for TM mode **a** Si waveguide (reprinted with permission from [10]. ©2015 Elsevier GmbH). **b** SU-8 waveguide

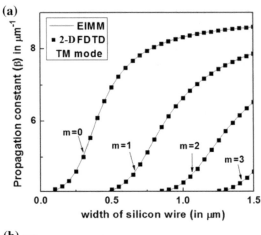

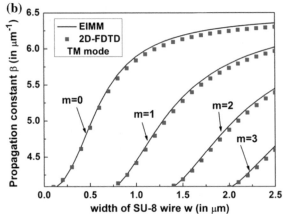

Fig. 2.10 Excitation efficiency versus propagation constant distribution of Si wire ($h = w = 0.25 \, \mu m$) for TE mode (reprinted with permission from [10]. ©2015 Elsevier GmbH)

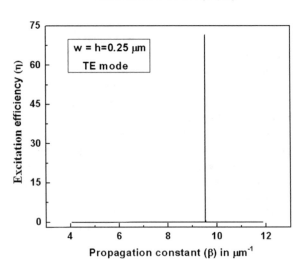

Fig. 2.11 Normalized mode
profile of Si photonic wire
with varying width for TE
and TM polarizations
(reprinted with permission
from [10]. ©2015 Elsevier
GmbH)

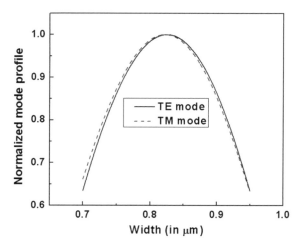

Fig. 2.12 Comparison of
normalized mode profile of
Si photonic wire between
EIMM and 2D-FDTD results
for TE mode (reprinted with
permission from [10]. ©2015
Elsevier GmbH)

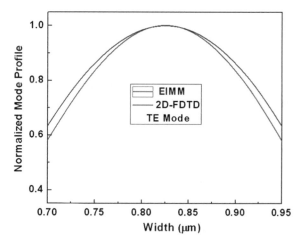

To estimate bending loss of circularly curved waveguides of bending radius R,
the equivalent lateral refractive index profile ($n_{eq}(u)$), as given in Eq. (2.28), was
computed (shown in Fig. 2.13).

To apply matrix method, n_{eq} was approximated by a number of step-index layers.
In the extreme right (1st layer in the computation), the n_{eq} was terminated beyond a
refractive index value equal to the peak refractive index of the structure. In some of
the computations, nonlinear layer thickness was considered to reduce computation
time. Typical variation of bending losses with respect to bending radii for Si and
SU-8 wires are shown in Figs. 2.14 and 2.15, respectively.

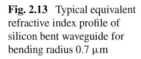

Fig. 2.13 Typical equivalent refractive index profile of silicon bent waveguide for bending radius 0.7 μm

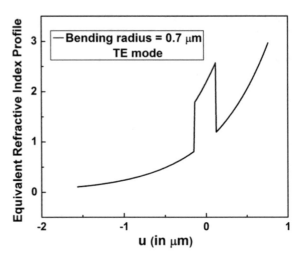

From our computed data, it can be noticed that bending loss increases exponentially with the decrease of bending radius. Since Si wires are all high-contrast waveguides, the estimated bending losses are negligibly small for bending radii greater than 1 μm as measured experimentally by Vlasov et al. [14].

Computed results in TE mode with slanted side wall angle of 35.26° for the Si wire waveguide is shown in Fig. 2.16. The top width and maximum thickness of the wire were taken as 250 nm each. From the results, we obtained showed that the structure still behaved as a single-mode waveguide with propagation constant 4.3421 μm^{-1}. We had also studied the wire with other inclination angles, viz. 15° and 45°, where the respective propagation constants were found to be 5.0638 and 4.1683 μm^{-1}. Thus, from the calculations, it can be seen that the propagation constant value of the guided modes decreases as the slant angle increases, i.e., reaching toward the cut-off value. This is obvious due to the reduction of effective contrast of waveguide when slant angle increases.

It is to note that to calculate propagation constants of graded refractive index structures, such as bent waveguides and wire waveguides with slanted walls, the number of layers considered (or transversal discretization) is an important parameter. For smaller layer thicknesses, EIMM yields more accurate results (approaching a limiting value). As an example, we had considered 321 layers for slanted wire waveguide of 35.26° inclination angle in order to obtain propagation constants accurately up to four decimal places. On increasing the number of layers beyond this, the value of β remained same up to four decimal places. On the other hand, for bending loss calculation, we had considered 3000 layers to compute Γ accurately up to around $\pm 2.3 \times 10^{-7} \mu m^{-1}$.

For single-mode silicon rib waveguide design, we had taken $H = 5$ μm and $h_s = 3.5$ μm. In the first step, effective index of the fundamental mode of the slab was computed by effective index method (EIM). It came around 3.4705 (propagation

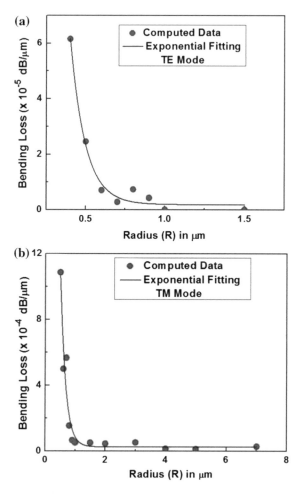

Fig. 2.14 Bending loss of bent silicon wire for different radii of curvature for **a** TE and **b** TM mode (reprinted with permission from [10]. ©2015 Elsevier GmbH)

constant was $14.0683\ \mu m^{-1}$). Thereafter, effective refractive indices of the rib region for the first two modes were computed. The computed values were 3.4737 and 3.4640, respectively. Since second-mode effective index at the rib region is less than effective index of fundamental mode at slab region, only one vertical mode will be guided in the rib region. Finally, transfer matrix method was applied on fundamental effective refractive index structure of this waveguide. The width of the rib was varied to determine the propagation constant values of the guided modes. The computed result for TE mode was validated with FDTD simulation and is shown in Fig. 2.17a. We may observe from this figure that the waveguide is single-moded within the rib width 0.9–6.2 μm for TE mode. It is to mention that the upper limit of the rib

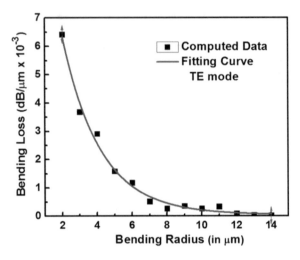

Fig. 2.15 Bending loss of bent SU-8 wire for different radii of curvature for TE polarization

Fig. 2.16 Excitation
efficiency versus variation of
propagation constant
distribution for silicon wire
waveguide with slanted wall
of 35.26° (reprinted with
permission from [10]. ©2015
Elsevier GmbH)

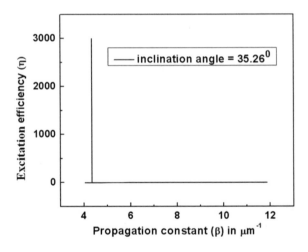

width for single-mode propagation as computed by EIMM is fairly close to the value
(6.4 μm) computed by Soref's condition [18] using the same design parameters.
Similar approach was adopted for TM mode with same H and h_s combinations and
we found that the waveguide supported single-mode within width 0.5–5.4 μm, which
is shown in Fig. 2.17b.

Fig. 2.17 EIMM and FDTD results for variation of propagation constant versus width for **a** TE_{00} and TE_{01} modes (reprinted with permission from [10]. ©2015 Elsevier GmbH). **b** TM_{00} and TM_{01} modes

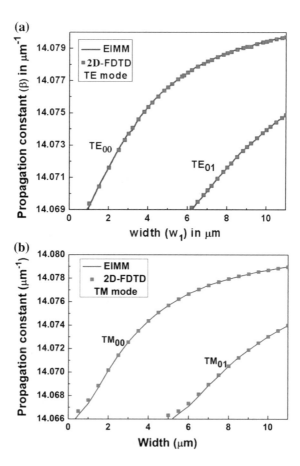

2.8 Conclusions

Design and analysis of silicon and SU-8 wire waveguides were done in this chapter where we had used effective index-based matrix method, which is less computation-intensive than other commercially available softwares such as FDTD or BPM. As an example, computation time of propagation constants by FDTD method of a wire waveguide took about ten times more than the computation time by EIMM in an Intel Core2 Duo PC. 2D-FDTD results of the structures were also studied here, which showed the accuracy of EIMM for these high refractive index contrast waveguides and applicability for both TE and TM polarizations. The method can also be extended for both silicon and SU-8 slot waveguides.

References

1. P. Ganguly, J.C. Biswas, S.K. Lahiri, Matrix based analytical model of critical coupling length of titanium in-diffused integrated-optic directional coupler on lithium niobate substrate. Fiber Integ. Opt. **17**, 139–155 (1998)
2. P. Ganguly, J.C. Biswas, S.K. lahiri, Analysis of titanium concentration and refractive index profiles of Ti:LiNbO$_3$ channel waveguide. J. Opt. **39**, 175–180 (2010)
3. R. Chakraborty, P. Ganguly, C. Biswas, S.K. Lahiri, Modal profiles in Ti:LiNbO3 two-waveguide and three-waveguide couplers by effective-index-based matrix method. Opt. Commun. **187**, 155–163 (2001)
4. T. Ghosh, B. Samanta, P.C. Jana, P. Ganguly, Comparison of calculated and measured refractive index profiles of continuous wave ultraviolate written waveguides in LiNbO$_3$ and its analysis by effective index based matrix method. J. Appl. Phys. **117**, 053106-1–7 (2015)
5. T. Ghosh, B. Samanta, P.C. Jana, P. Ganguly, Design of a directional coupler based on UV-induced LiNbO$_3$ waveguides. J. Opt. Commun. **38**, 255–262 (2017)
6. K.S. Chiang, Analysis of optical fibers by the effective-index method. Appl. Opt. **25**, 348–354 (1986)
7. A. Ghatak, K. Thyagarajan, M.R. Shenoy, Numerical analysis of planar optical waveguides using matrix approach. J. Lightwave Technol. **5**, 660–667 (1987)
8. M.R. Shenoy, K. Thyagarajan, A. Ghatak, Numerical analysis of optical fibers using matrix approach. J. Lightwave Technol. **6**, 1285–1291 (1988)
9. M.R. Ramadas, R.K. Varshney, K. Thyagarajan, A.K. Ghatak, A matrix approach to study the propagation characteristics of a general nonlinear planar waveguide. J. Lightwave Technol. **7**, 1901–1905 (1989)
10. S. Samanta, P. Banerji, P. Ganguly, Effective index-based matrix method for silicon waveguides in SOI platform, Optik. Int. J. Light Electron Opt. **126**, 5488–5495 (2015)
11. M. Heiblum, Analysis of curved optical waveguides by conformal transformation. IEEE J. Quantum Electron. **11**, 75–83 (1975)
12. C.K. Tang, G.T. Reed, A.J. Walton, A.G. Rickman, Low loss and single mode phase modulator in SIMOX material. J. Lightwave Technol. **12**, 1394–1400 (1994)
13. Q. Weiping, F. Dagang, Modal analysis of a rib waveguide with trapezoidal cross section by variable transformed Galerkin method, in *Proceedings of the International Conference on Computational Electromagnetics and its Applications* (1999), pp. 82–85
14. Y.A. Vlasov, S.J. McNab, Losses in single-mode silicon-on-insulator strip waveguides and bends. Opt. Express **12**, 1622–1631 (2004)
15. E.D. Palik, *Handbook of Optical Constants of Solids* (Academic Press, Boston, 1985)
16. D.E. Aspnes, J.B. Theeten, Spectroscopic analysis of the interface between si and its thermally grown oxide. J. Electrochem. Soc. **127**, 1359–1365 (1980)
17. B. Yang, Y. Zhu, Y. Jiao, L. Yang, Z. Sheng, S. He, D. Dai, Compact arrayed waveguide grating devices based on small SU-8 strip waveguides. J. Lightwave Technol. **29**, 2009–2014 (2011)
18. R.A. Soref, J. Schmidtchen, K. Petermann, Large single-mode rib waveguides in GeSi-Si and Si-on-SiO/sub 2. IEEE J. Sel. Top. Quantum Electron. **27**, 1971–1974 (1991)

Chapter 3
Experimental Studies on SU-8 Wire Waveguides

3.1 Introduction

Optical waveguides made of SU-8 polymer are increasingly employed to fabricate passive integrated optic devices, such as gratings, optical filters, and microoptical sensors based on Mach–Zehnder Interferometer, optofluidic, and micro-opto-electro-mechanical systems (MOEMS) [1–10]. As stated in Chap. 1, SU-8 has a wide transparency region in visible and near-infrared wavelength of light, which made this polymer a suitable waveguide material for a variety of applications. In general, SU-8 waveguides are fabricated by using I-line (365 nm) photolithography [1, 11], although some researchers [12–16] have also used maskless direct laser writing technique to fabricate SU-8 waveguides. Schroder et al. [12] presented the simulation and first experimental implementation of a novel polymer 3-D waveguide for on-chip communication. They used direct laser writing of 780 nm wavelength, and 70 fs pulse width for 3-D laser lithography. Feinaeugh et al. [13] had used laser-induced backward transfer using a femtosecond laser to write SU-8 waveguides. Parida et al. [14] optimized the parameters to fabricate optical components like photonic bandgap structures, splitters, directional couplers, and gratings with these SU-8 waveguides by laser writing and electron beam lithography. By introducing H-nu 470 photoinitiator, Ramirez et al. [15] shifted the absorption peak of SU-8 from 365 to 470 nm, and with this modified SU-8 they produced single- and multi-mode waveguides at 405 nm by direct writing technique. Sum et al. [16] used high-energy proton submicron beam to directly write on SU-8 resists and fabricated low-loss SU-8 channel waveguides.

This chapter presents experimental methods to study the fabrication and characterization of multi/single-mode SU-8 wire waveguides developed in author's laboratory [17–21]. Air-cladded and polydimethylsiloxane (PDMS)-cladded SU-8 waveguides were fabricated by continuous-wave laser direct writing process at 375 nm writing wavelength. Fiber–waveguide coupling loss and propagation loss of fabricated waveguides were measured by cutback method. The fundamental mode for the

© The Author(s), under exclusive license to Springer Nature Singapore Pte Ltd. 2020
S. Samanta et al., *Photonic Waveguide Components on Silicon Substrate*,
SpringerBriefs in Applied Sciences and Technology,
https://doi.org/10.1007/978-981-15-1311-4_3

waveguide was excited by precise fiber positioning, and mode image was recorded. The mode profiles, mode indices, and refractive index profiles were extracted from this mode image of the fundamental mode which matched remarkably well with the theoretical predictions using effective index-based matrix method. At the end, a feasibility study was also made to fabricate SU-8 waveguides by focused ion beam (FIB) lithography; however, we found that this technique was not practically suitable for fabricating long waveguide structures.

3.2 Fabrication and Characterization

3.2.1 Fabrication by Laser Direct Writing Technique

We started the fabrication process with cleaning of (100) silicon wafer by heat treatment with acetone around 50 °C for 10 min in order to remove organic residues from the wafer surface. The removal of organic material continued with piranha cleaning, which was made up of 1:1 volume ratio of concentrated sulphuric acid (H_2SO_4, 98%) and hydrogen peroxide (H_2O_2). This solution being extremely exothermic started to bubble and heated up and was prepared by pouring concentrated H_2SO_4 slowly into H_2O_2 [22]. Then this mixture was kept unaltered for about 25 min; thereafter, the wafer was taken out and was rinsed thoroughly with DI (deionized) water having resistivity 18.2 MΩ cm and dried with nitrogen jet. Thermal oxidation of this silicon wafer was done next where the sample was put inside the oxidation furnace (Tempress Systems diffusion furnace, the Netherlands) and waited till temperature reached to 1050 °C (additional 10 min wait for stabilization); the bubbler temperature being around 97 °C. In order to establish a good Si–SiO_2 interface (i.e., for improving the adhesion), dry oxidation was used first. But due to the lower growth rate, dry oxidation could not be used for long term, so for thicker layers, wet oxidation was used. For wet oxidation, water (H_2O) was used which dissociated at high temperatures to form hydroxide that diffuse faster in silicon compared to molecular oxygen (O_2) of dry oxidation [23]. Again, the top oxide surface should be of good quality, and thus, dry oxidation was used for the top surface. So, the total thermal oxidation process was done in a sequence of dry–wet–dry for 30 min–2 h–30 min at a temperature of 1050 °C. After cooling down the temperature to 700 °C, the sample was taken out and the thickness of this formed SiO_2 layer was measured by using a single wavelength (632.8 nm, He–Ne laser) ellipsometer (L116E Gaertner, USA). It measured the change of polarization upon reflection from the sample. The polarization change was quantified by the amplitude ratio (AR) and the phase difference (Δ). Since the reflected signal depends on thickness as well as the material properties, ellipsometry can be used to measure thickness or refractive index or both of a transparent thin film in a non-contact manner. In our experiment, we had supplied the trial SiO_2 refractive index and thickness values initially, and the controlling software of the machine followed an iterative procedure to calculate AR and Δ using Fresnel equations. The best-matched computed AR and Δ values with the experimental data provide the thickness and refractive index of the SiO_2 layer. The measured thickness and refractive index of SiO_2 film were ~1.0 μm and 1.46, respectively, at 632.8 nm wavelength, which was fairly uniform throughout the wafer surface. Next, it was treated with a general heating for 30 min at a temperature of 150 °C and cooled down to room temperature. Oxygen plasma treatment (Zepto, Diener Electronic, Germany) was carried out (to improve the adhesion of SU-8 on SiO_2) at a power of 40 W for 45 s. Following this, negative photoresist SU-8 2005 (viscosity 290 cSt) [24] was spin-coated on the sample at 500 rpm for 10 s and then ramped up to 6000 rpm for next 20 s to achieve a resist thickness of ~1.35 μm, which was verified by using a surface profiler (Dektak 150, Veeco Sloan). The sample was then baked on a horizontal hot plate at 65 and 95 °C for 1 and 3 min, respectively, for solvent evaporation. It was then allowed to cool down to room temperature for 10–15 min for thermal relaxation and thereafter was placed in a direct laser writing system (Microtech laser writer, LW-2002). A 375 nm UV laser source, fitted with the laser writing system, was used to obtain the waveguide pattern of different widths parallel to (110) plane of the substrate. Single scanning (speed: 320 μm/s) of the laser beam (898 mJ/cm^2) was used in our waveguide fabrication process. Post-baking of the substrate on the hot plate was then carried out for 1 min at 65 °C and next 1 min for 95 °C for proper cross-linking of the polymer. Next, it was dipped into Microchem SU-8 developer and carefully developed for 15 s. Finally, it was rinsed with isopropyl alcohol for complete development verification. The waveguide writing process was carefully optimized by varying the laser beam intensity, scan speed, and development time. For some fabricated SU-8 waveguides, low-index polydimethylsiloxane (PDMS) was used as superstrate instead of air. For those cases, after fabricating SU-8 wire waveguides, samples were processed with an additional step of PDMS coating. Here, first of all 10:1 volume ratio of PDMS and curing agent were mixed in a beaker, and the mixture was stirred until bubble appears. Then the beaker was kept in a desiccator connected with a vacuum pump until all bubble disappears. The sample was then spin-coated with this mixture at 500 rpm for 10 s and ramped to 8000 rpm for next 20 s in order to obtain a 10-μm-thick PDMS layer. A final heat treatment was done in hot plate at 95 °C for 30 min.

3.2.2 Characterization

The fabricated waveguide samples were manually cleaved by applying pressure with a sharp tweezer at 90° angle to the waveguides. The waveguide structures were inspected under optical microscope (Union, SG-V 84149) and scanning electron microscope (Merlin Zeiss SEM), and the widths of different waveguides were measured. It was noticed that the minimum width that we could achieve by our direct writing process was around 5.3 μm, which was expected since the UV laser beam coming out from the Microtech laser writer was ~5.0 μm in diameter. The photomicrographs of the fabricated SU-8 waveguide of 1.35 μm height and 5.3 μm width, with its cleaved edge, are shown in Fig. 3.1a, b.

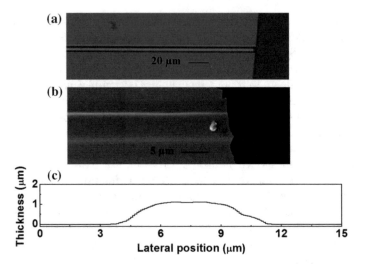

Fig. 3.1 Waveguide edge **a** Optical microscopic view; **b** SEM view; **c** Cross-sectional profile of SU-8 waveguide (reprinted with permission from [21]. ©2016 Elsevier B.V.)

Next, we had scanned the SU-8 wire waveguide by the surface profiler in order to obtain the cross-sectional profile of the waveguide. A typical profile of the waveguide is presented in Fig. 3.1c. As the laser beam output had a Gaussian intensity distribution, the cross-sectional profile of the waveguide was not exactly rectangular, but trapezoidal in shape having full width at half maximum (FWHM) ~5.3 µm.

3.2.2.1 Loss Measurement of Air- and PDMS-Cladded Waveguide

The insertion losses of the cleaved waveguides of different lengths were measured by using the setup shown in Fig. 3.2. A semiconductor laser source emitting unpolarized 1.55 µm wavelength of ~1mW power (CVI Melles Griot, USA) was converted to TE polarized light by using fiber polarizer and carefully coupled to the cleaved waveguide

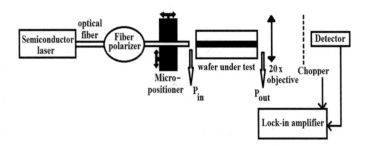

Fig. 3.2 Set-up for loss measurement (reprinted with permission from [21]. ©2016 Elsevier B.V.)

samples of different lengths using high-precision five-axis micropositioner (ULTRA-lign, Model: 561D, Newport). The waveguide output was recorded by a fixed frequency chopper–InGaAs detector (Model: 71905, Oriel Instruments)–lock-in amplifier (Model: SR 830 DSP, Stanford Research Systems) arrangement to improve the signal-to-noise ratio (SNR). The chopping frequency was 11.35 Hz. Then after removing the sample, the fiber output was also recorded. The total insertion losses (in dB) for waveguides of different lengths were measured from the relation: $-10\log_{10}$ (P_{in}/P_{out}), where P_{in} and P_{out} were the fiber output and waveguide output, respectively. The measured insertion losses versus lengths of the waveguides were plotted and best-fitted with a straight line. The average propagation loss (in dB/mm) of the fabricated waveguide was estimated from the slope of this straight line and its intercept on the vertical axis through origin indicated average coupling loss. The coupling loss included loss from fiber-to-waveguide and loss from waveguide-to-$20\times$ objective lens. The measured propagation loss for our fabricated air-cladded SU-8 waveguide was found to be 0.51 dB/mm, and that of PDMS-cladded waveguide was measured as 0.30 dB/mm. This air-cladded loss value of 0.51 dB/mm at 1550 nm wavelength matched well with previously reported results by Ramirez et al. [15], where the propagation loss was 0.44 dB/mm at 633 nm transmitting wavelength. The values of coupling losses for clad as air and PDMS were 2.18 and 4.11 dB, respectively. Figure 3.3 shows the measured insertion losses for different waveguide lengths with PDMS cladding.

Since waveguide edge preparation plays an important role to reduce coupling loss of the waveguides (which require the waveguide end-faces to be optically flat and free of defects to a very high degree), we also studied the effect of edge-polishing of these polymer waveguides. For this, a PDMS-cladded SU-8 waveguide of length 13.5 mm was sandwiched in between two glass substrates and pressed together to protect the SU-8 waveguide ends from any possible damage during polishing; and edge-polished it using a polishing machine (6390.1, Ultratech Inc., USA) with different

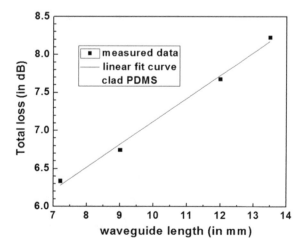

Fig. 3.3 Total loss versus different waveguide lengths (reprinted with permission from [21]. ©2016 Elsevier B.V.)

pressures, rotation speeds (10–20 rpm), and diamond plates (0.1–6.0 μm). In between the polishing steps, the waveguide edge was observed under an optical microscope and mild ultrasonic cleaning was done to remove contamination of higher order particles. After precise polishing of both the edges of the waveguide, the insertion loss of this waveguide was again measured. Since our polishing speed and applied pressure were very low, and also we had used water-soluble coolant during the entire polishing process, we did not observe any adverse effect, such as melting, or peeling-off of the SU-8 film from the substrate. The measured total insertion loss of the polished waveguide of length 13.5 mm was around 6.28 dB, which was ~8.23 dB before polishing. Thus, the coupling loss of 4.11 dB (before polishing) had reduced to 2.23 dB after polishing.

3.2.2.2 Mode Output of Air- and PDMS-Cladded Waveguide

To observe the mode profiles of the fabricated waveguides, an experimental setup as shown in Fig. 3.4 was used. The images of the waveguide outputs were recorded using an IR camera (Electrophysics Micronviewer, Model: 7290A) and were observed in an ULTRAK monitor.

The recorded near-field image of the air-cladded SU-8 waveguide having width 5.3 μm and height 1.35 μm is shown in Fig. 3.5. From this image, we can see that the waveguide is strongly multi-mode in nature laterally. After coating with PDMS, the number of modes was reduced to a great extent, which can be seen from the mode image of the PDMS-cladded waveguides (Fig. 3.6a). Now, the existence of the number of guided modes at the output of the waveguide of certain length depends on coupling conditions, i.e., the position and inclination of the input fiber with respect

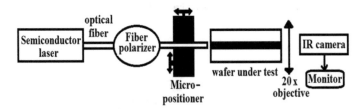

Fig. 3.4 Set-up for mode image recording (reprinted with permission from [21]. ©2016 Elsevier B.V.)

Fig. 3.5 Near-field image of air-cladded SU-8 waveguide (reprinted with permission from [21]. ©2016 Elsevier B.V.)

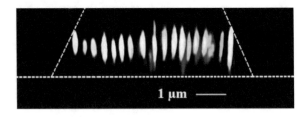

Fig. 3.6 **a** Near-field image of PDMS-cladded multi-mode SU-8 waveguide (reprinted with permission from [21]. ©2016 Elsevier B.V.). **b** Near-field image of PDMS-cladded single-mode SU-8 waveguide (reprinted with permission from [21]. ©2016 Elsevier B.V.)

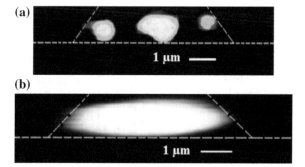

to the waveguide input edge. After precise adjustment of fiber–waveguide coupling in the experimental setup used (Fig. 3.4), we were able to capture the mode image of isolated fundamental mode for 13.5-mm-long PDMS-cladded waveguide. This is shown in Fig. 3.6b. Approximate outlines of the SU-8 waveguide are included in Figs. 3.5 and 3.6. All the optical measurements were performed on a vibration isolation table (Holmarc, India).

3.2.2.3 Lateral Mode Profile of PDMS-Cladded Waveguide

The near-field image of the fundamental mode of the waveguide at 1.55 μm transmitting wavelength was image processed using MATLAB programming in order to obtain the mode profiles (both horizontal and vertical) of the waveguide. Figure 3.7 shows the normalized lateral and depth intensity profiles of the PDMS-cladded SU-8 waveguide obtained from mode imaging of fundamental mode (Fig. 3.6b). The recorded intensity of the waveguide mode images (Figs. 3.5 and 3.6) was kept just above the saturation limit of the IR camera for a better visual perspective. Whereas, the recorded mode profile displayed in Fig. 3.7 was extracted after attenuating the waveguide output (Fig. 3.6b) through an adjustable attenuator such that the maximum intensity was below the saturation limit of the camera. The magnification of the experimental setup was carefully measured. The Gaussian nature of the experimental intensity profiles of the waveguide indicated the single-mode behavior. It is to mention that the vertical mode profile of Fig. 3.7b is a bit asymmetric at the ends; the reason is due to the difference in refractive index profiles of the SiO_2-SU-8-PDMS interfaces. Whereas, for the lateral mode profile, as both the sides are having PDMS and there is SU-8 in between, the profile is nearly symmetric.

3.2.2.4 Refractive Index Profile of PDMS-Cladded Waveguide

From the obtained fundamental mode intensity profiles of the PDMS-cladded polymer waveguide, the refractive index profiles in lateral and depth directions were estimated using scalar wave equation:

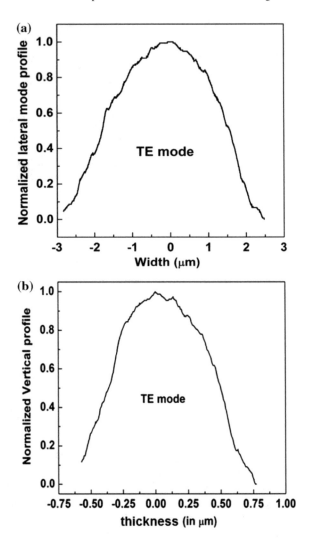

Fig. 3.7 Normalized **a** lateral, **b** vertical mode profile as obtained from mode imaging for PDMS-cladded SU-8 waveguide for TE polarization

$$\nabla^2 \Psi + (k^2 n^2 - \beta^2)\Psi = 0 \tag{3.1}$$

where Ψ is the electric field; n is the refractive index; k being the wave vector; and β is propagation constant. Now, Eq. (3.1) can be written as follows:

$$n^2 = \frac{\beta^2}{k^2} - \frac{1}{k^2} \frac{\nabla^2 \Psi}{\Psi} \tag{3.2}$$

If n_s be the substrate refractive index and n_{effo} the effective refractive index or mode index of the fundamental mode of the waveguide, then refractive index change Δn can be expressed as below:

$$\Delta n = \sqrt{\frac{\beta^2}{k^2} - \frac{1}{k^2}\frac{\nabla^2\Psi}{\Psi}} - n_s$$

$$\text{or,} \quad \Delta n = \sqrt{n_{effo}^2 - \frac{1}{k^2}\frac{\nabla^2\Psi}{\Psi}} - n_s \tag{3.3}$$

The flowchart of the total computation process is shown in Fig. 3.8. The data of mode intensity profile was taken as input from which electric field (Ψ) was determined, which is related as square root of the measured mode intensity profile. Next, fast Fourier transform (FFT) of Ψ was computed and was smoothed by using a third-order low-pass Butterworth filter with the transfer function in s plane $H(s)$ and amplitude response function $|H(j\omega)|$ as follows [25, 26]:

$$H(s) = \frac{1}{(s+1)(s^2+s+1)}; \quad s = \frac{j\omega}{\omega_c}$$

$$|H(j\omega)| = \frac{1}{\sqrt{1 + \varepsilon^2\left(\frac{\omega}{\omega_c}\right)^6}};$$

$$\omega_c = \text{3-dB cut-off frequency, and } \varepsilon^2 = 0.99526231 \tag{3.4}$$

Fig. 3.8 Flowchart of refractive index change computation

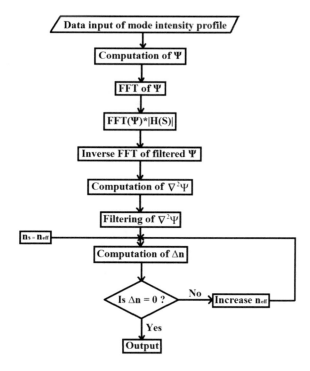

The same Butterworth filter was used for $\nabla^2 \Psi$ determination in order to remove the high-frequency noise. It may be mentioned that the choice of the third-order Butterworth filter was due to its flat frequency response and smooth attenuation of high frequencies without fully eliminating them [27]. Now, as the refractive index change is a small quantity, initially we had considered mode index equal to the substrate refractive index value, and Δn was computed. In the next step, value of n_{effo} was increased slowly until the minimum refractive index change became zero, which gave the exact mode index value.

The estimated horizontal and vertical refractive index distributions of the waveguide extracted from the near-field image are shown in Fig. 3.9. The cut-off frequencies of the Butterworth filter, for both the horizontal and vertical profiles, were chosen to have above 98% transmitted power. n_{effo} values were found to be 1.527 and 1.532 from respective horizontal and vertical refractive index profiles, which were in good agreement with the computed value (1.5263) using EIMM. It may also be noted from these figures that the horizontal profile was nearly symmetrical having an index contrast of 0.1435; this was very near to the expected theoretical value using EIMM (0.1316), which was calculated from the difference of $n_{\text{eff}}|_{h=1.35 \, \mu m}$ value (1.5316) and PDMS refractive index value (1.4). The vertical profile, on the other hand, was

Fig. 3.9 a Horizontal.
b Vertical distance versus refractive index change for PDMS-cladded waveguide (reprinted with permission from [21]. ©2016 Elsevier B.V.)

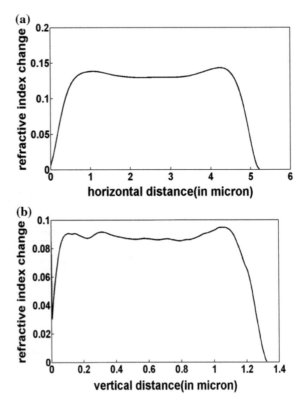

slightly asymmetric, and index contrast obtained from this profile was 0.0947, which was again in good agreement with the expected value (0.1216). The asymmetry in index change values (0.031) between two sides of the obtained vertical distribution indicated the index difference between SiO_2 and PDMS layer (0.047).

3.2.2.5 Comparison Between Fabricated and Theoretical Results

Figure 3.10 shows the plot of effective refractive indices with varying film thicknesses as obtained using effective index-based matrix method, which indicates that the waveguide operates in single-mode region within film thickness (h) of $0.43 \leq h$ (μm) ≤ 1.72 for air cladding and $0.23 \leq h$ (μm) ≤ 1.52 for PDMS cladding. Now, as our fabricated waveguide thickness was 1.35 μm, these are supporting a well-confined fundamental mode in the depth direction, for both the superstrates. The excitation efficiency versus propagation constant plot for waveguide thickness of 1.35 μm and width 5.3 μm is shown in Fig. 3.11. It can be seen from Fig. 3.11a that the waveguide with clad as air is highly multi-mode in nature, having a number of resonant peaks. The PDMS-cladded waveguide of the same dimensions, on the other hand, supports lesser number of modes. The experimental results of Figs. 3.5 and 3.6a match with these theoretical results as obtained by EIMM. The effective refractive index or mode index values of the fundamental mode of the waveguide are found to be 1.527 and 1.532 from respective horizontal and vertical refractive index profiles, which are in good agreement with the computed value (1.5263) using EIMM. It may be also be noted from these figures that the horizontal profile is nearly symmetric having an index contrast of 0.1435 which is very near to the expected theoretical value (0.1316). Also, the computed lateral mode profile matches fairly well with the obtained experimental profile (Fig. 3.12). Slightly higher mode width

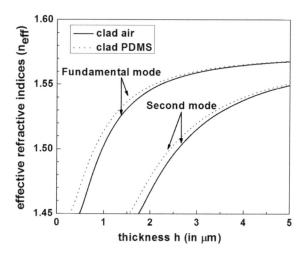

Fig. 3.10 Effective refractive indices for different film thickness (reprinted with permission from [21]. ©2016 Elsevier B.V.)

Fig. 3.11 Excitation
efficiency versus propagation
constant **a** clad air, **b** clad
PDMS (reprinted with
permission from [21]. ©2016
Elsevier B.V.)

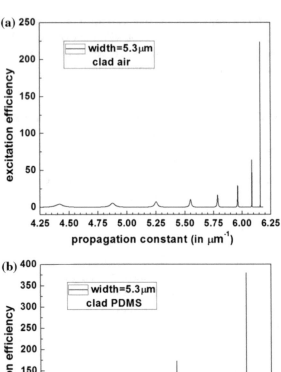

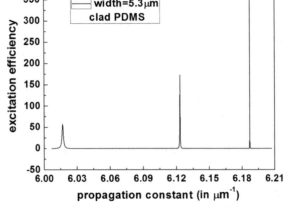

and asymmetry of the measured profile compared to the theoretical one attribute to little defocusing and misalignment associated during imaging.

3.2.2.6   Discussions

The minimum spot size of the laser output beam of our direct laser writing system (LW-2002) was ~5 μm, so we had achieved a minimum waveguide width of ~5.3 μm by laser direct writing process, which supported multi-mode operation of the air-cladded and PDMS-cladded SU-8 waveguides. However, on changing the position and inclination of the input coupling fiber with respect to the waveguide input edge, only fundamental mode of the waveguide had been excited. The experimental characterization results for the fundamental mode of the PDMS-cladded

Fig. 3.12 Comparison of normalized lateral mode profile between experimental data and effective index-based matrix method for PDMS-cladded SU-8 waveguide (reprinted with permission from [21]. ©2016 Elsevier B.V.)

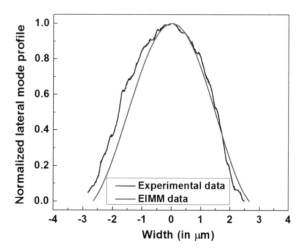

SU-8 waveguide matched remarkably well with the theoretical expectations. The refractive change profiles along vertical and lateral directions were extracted from the measured mode profiles of single-mode waveguide. The process used for this extraction is applicable for slowly varying graded-index or low-contrast step-index waveguides. The computation of mode profile using numerical or semi-analytical EIMM technique was also considered and compared with the experimental profile as shown in Fig. 3.12. A reasonably good agreement of the results indicates the validity of EIMM technique. The waveguide propagation loss of these waveguides was measured by using a modified cut-back technique. The measured propagation losses of air- and PDMS-cladded SU-8 waveguides were in good agreement with the reported data of other researchers [12, 28–30]. It may be noted that all theoretical and experimental results presented so far in this chapter were for TE polarization of light for which light confinement was better than TM modes. Since to fabricate a strictly single-mode air-cladded waveguide, one has to obtain waveguide width less than a micron, we made an experimental study with focused ion beam (FIB) lithography with an aim to have optical waveguides of lesser width. Section 3.2.3 discusses our experimental study regarding the same.

3.2.3 Fabrication by Focused Ion Beam Lithography

Figure 3.13 shows the flow diagram of sample preparation and fabrication of waveguides. Up to step-6 of Fig. 3.13, the procedures are same as stated in Sect. 3.2.1. Thereafter, the sample was placed on Quorum DC sputtering gold–palladium (80:20%) coater for about 2 min before proceeding to focused ion beam (FIB) milling (using AURIGA Compact Zeiss FIB machine). A high resolution of 30 keV Ga^+ ions in combination with a high-precision field emission gun was used for proper positioning and

Fig. 3.13 Flow diagram of
sample preparation and
fabrication

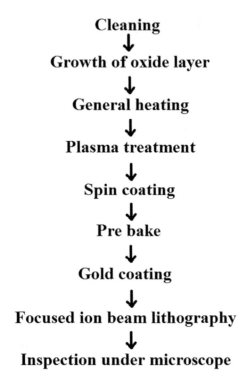

Cleaning
↓
Growth of oxide layer
↓
General heating
↓
Plasma treatment
↓
Spin coating
↓
Pre bake
↓
Gold coating
↓
Focused ion beam lithography
↓
Inspection under microscope

inspection of the fabricated structures. The pattern drawn in standard GDSII format
(using Elphy Quantum software, Raith) consisted of five parallel straight waveguides
of varying widths and was written directly on the wafer surface with a pattern gen-
erator; the writing method being raster scanning in dot sequence. Figure 3.14 shows
the description method for FIB apparatus. The stage with sample was tilted 54° so
that it becomes perpendicular to the ion gun, which was necessary for ion milling.
The working distance for simultaneously milling and imaging (coincidence point)
was ~5 μm.

Fig. 3.14 Description
method for FIB apparatus
(reprinted with permission
from [20]. ©2017
Cambridge University press,
Materials Research Society)

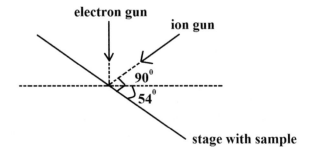

3.2.4 Characterization

The maximum writing field for AURIGA Compact FIB machine was 197 μm^2, beyond which area-stitching was required. Also, the minimum spot size that we could get was 30 nm with a minimum current of 1 pA. In order to make SU-8 applicable for waveguide purpose, that means, for proper optical characterization, patterns should be fabricated edge-to-edge of the wafer. Moreover, for easy handling of sample, the sample length should be at least 3–4 mm. Now, to scan a sample of 3–4 mm length with small ion dose would take a huge amount of time, which was practically not feasible. Due to the heavy mass of Ga^+ ions, interaction of gallium beam with the sample surface was destructive in nature. Also, high-energy ion bombardment on the sample surface might gradually gather several volts of charge resulting in cavity or hole, and local melting due to electrostatic discharge. Looking at the limitations, we had chosen a beam current of 10 nA, which was kept fixed (beam shape being circular and a beam diameter or spot size of 1.5 μm), and ion doses were varied. The milling depth was adjusted by this ion dose (i.e., the time for which the ion stayed on each spot while milling). The widths of the waveguides after fabrication as inspected under FESEM did not match as drawn in the CAD pattern. The widths were broadened due to the application of high beam current of 10 nA (shown in Fig. 3.15), as FIB has a drawback of backscattering and re-deposition while milling of structures. Actually, when we had increased the ion dose to achieve more depth, the beam speed decreased, which automatically resulted into broader width.

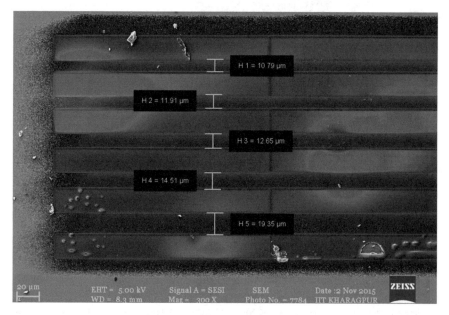

Fig. 3.15 Fabricated waveguides of different widths (reprinted with permission from [20]. ©2017 Cambridge University press, Materials Research Society)

The milling depths as obtained from atomic force microscopy (AFM) for 1, 15, and 40 C/cm^2 ion doses were 0.195 μm, 0.34 μm, and 0.57 μm, respectively (as shown in Table 3.1). The root-mean-square (rms) roughness values were 0.118 μm, 0.415 μm, and 0.512 μm for 1, 15, and 40 C/cm^2 ion doses, respectively.

As a thumb rule, the permissible roughness for waveguide fabrication is one-tenth of the transmitting wavelength (1.55 μm), above which the scattering loss will be high; thus, ion dose 1 C/cm^2 with roughness 0.118 μm is only acceptable for waveguide writing purpose. However, the milling depth that we had achieved with this ion dose 1 C/cm^2 was only 0.195 μm. In order to obtain more depth, the ion dose had to be increased; but at the same time, there was an increase in the surface roughness as well, which meant more prone to optical losses. Thus, our study on waveguide writing process using FIB method reveals that this method is practically not suitable for fabricating long SU-8 waveguide structures, although it may be

Table 3.1 Topography of waveguides for different ion doses and corresponding milling depth as obtained using atomic force microscope

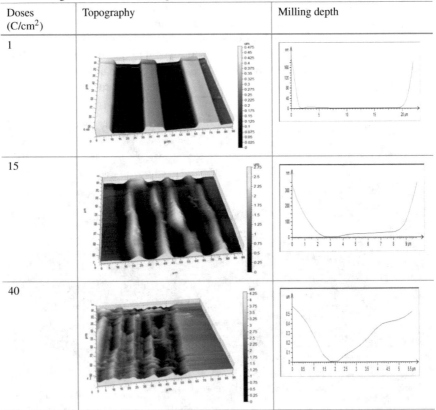

Doses (C/cm^2)	Topography	Milling depth
1		
15		
40		

Reprinted with permission from [20]. ©2017 Cambridge University press, Materials Research Society

Fig. 3.16 SEM view of **a** two-waveguide optical coupler as obtained by FIB lithography, **b** inset showing the separation of 0.15 μm in between coupled waveguides (reprinted with permission from [20]. ©2017 Cambridge University press, Materials Research Society)

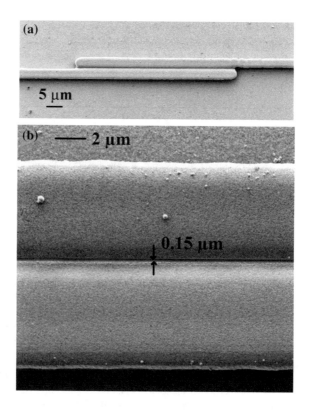

suitable for microstructuring along or over photonic waveguide structures within a very small region. As an example, two-waveguide optical coupler with 0.15 μm separation between coupled waveguides fabricated using FIB lithography is shown in Fig. 3.16.

3.3 Conclusions

We had experimentally studied SU-8 wire waveguides fabricated on oxidized silicon substrate. Fabrication attempts were made using direct laser writing technique at 375 nm writing wavelength and focused ion beam lithography. Minimum width of waveguide that we achieved by laser writing process was 5.3 μm. These SU-8 waveguides with air- and PDMS cladding had propagation losses of 0.51 dB/mm and 0.30 dB/mm, respectively. Mode index and refractive index profiles of the PDMS-cladded waveguides were extracted from measured fundamental mode profile at 1550 nm transmitting wavelength for TE polarization. While processing with FIB lithography, we found that although it was an effective technique in rapid fabrication of several prototype devices of very small footprint, it was not favorable for fabrication of long conventional SU-8 waveguide structures. However, it might be well-suited in fabricating photonic crystal structures or making any precise modifications in micro- and nanometer photonic waveguide structures.

References

1. M. Nordstrom, D.A. Zauner, A. Boisen, J. Hubner, Single-mode waveguides with SU-8 polymer core and cladding for MOEMS applications. J. Lightwave Technol. **25**, 1284–1289 (2007)
2. C.S. Huang, W.C. Wang, SU8 inverted-rib waveguide Bragg grating filter. Appl. Optics **52**, 5545–5551 (2013)
3. S.Q. Xie, J. Wan, B.R. Lu, Y. Sun, Y. Chen, X.P. Qu, R. Liu, A nanoimprint lithography for fabricating SU-8 gratings for near-infrared to deep-UV application. Microelectron. Eng. **85**, 914–917 (2008)
4. X. Shang, Y. Tian, M.J. Lancaster, S. Singh, A SU8 micromachined WR-1.5 band waveguide filter. IEEE Microw. Wireless Compon. Lett. **23**, 300–302 (2013)
5. N. Pelletier, B. Beche, N. Tahani, L. Camberlein, E. Gaviot, A. Goullet, J.P. Landesman, J. Zyss, Integrated Mach-Zehnder interferometer on SU-8 polymer for designing pressure sensors, in *IEEE Sensors* (2005), Irvine, CA, pp. 640–643
6. M. Bednorz, M. Urbanczyk, T. Pustelny, A. Piotrowska, E. Papis, Z. Sidor, E. Kaminska, Application of SU8 polymer in waveguide interferometer ammonia sensor. Mol. Quantum Acoust. **27**, 31–40 (2006)
7. B. Sepulveda, J.S. Rio, M. Moreno, F.J. Blanco, K. Mayora, C. Dominguez, L.M. Lechuga, Optical biosensor microsystems based on the integration of highly sensitive Mach-Zehnder interferometer devices. J. Optics A. Pure Appl. Opt. **8**, S561–S566 (2006)
8. C. Prokopa, N. Irmler, B. Laegel, S. Wolff, A. Mitchell, C. Karnutscha, Optofluidic refractive index sensor based on air-suspended SU-8 grating couplers. Sens. Actuators, A **263**, 439–444 (2017)
9. I.C. Liu, P.C. Chen, L.K. Chau, G.E. Chang, Optofluidic refractive-index sensors employing bent waveguide structures for low-cost, rapid chemical and biomedical sensing. Opt. Exp. **26**, 273–283 (2018)
10. V. Anvekar, T. Kundu, S. Mukherji, Gold capped SU-8 nanoridges as plasmonic sensor, in *International Conference on Optics and Photonics* (2013), Zhongli, Taiwan
11. B. Beche, N. Pelletier, E. Gaviot, J. Zyss, Single-mode TE_{00}–TM_{00} optical waveguides on SU-8 polymer. Opt. Comm. **230**, 91–94 (2004)
12. M. Schroder, M. Bulters, C.V. Kopylow, R.B. Bergmann, Novel concept for three-dimensional polymer waveguides for optical on-chip interconnects. J. Eur. Opt. Soc. Rap. Pub. **7**, 12027 (2012)
13. M. Feinaeugle, D.J. Heath, B. Mills, J.A. Grant-Jacob, G.Z. Mashanovich, R.W. Eason, Laser-induced background transfer of nanoimprinted polymer elements. Appl. Phys. A **122**, 398 (2016)
14. O.P. Parida, N. Bhatt, Characterization of optical properties of SU-8 and fabrication of optical components, in *ICOP—2009 International Conference on Optics and Photonics*, Chandigarh, India (2009)
15. J.C. Ramirez, J.N. Schianti, M.G. Almeida, A. Pavani, R.R. Panepucci, H.E. Hernandez-Figueroa, L.H. Gabrielli, Low-loss modified SU-8 waveguides by direct laser writing at 405 nm. Opt. Mat. Exp. **7**, 2651–2659 (2017)
16. T.C. Sum, A.A. Bettiol, J.A.V. Kan, F. Watt, E.Y.B. Pun, K.K. Tung, Proton beam writing of low-loss polymer optical waveguides. Appl. Phys. Lett. **83**, 1707–1709 (2003)

17. P.K. Dey, S. Samanta, P. Ganguly, Fabrication of ridge polymer waveguide by direct laser writing at 375 nm wavelength, in *12th International Conference on Fiber Optics and Photonics* (2017), IIT Kharagpur, India, M2B.6

18. S. Samanta, P.K. Dey, P. Banerji, P. Ganguly, Fabrication of SU-8 wire waveguide on silicon substrate, in *International Conference on Light Quanta: Modern Perspectives and Applications* (2015), Allahabad, India, S5A.53

19. S. Samanta, P.K. Dey, P. Banerji, P. Ganguly, Fabrication of SU-8 polymer waveguides using focused ion beam lithography, in *MRS Fall Meeting & Exhibit* (2016), Boston, USA, PM1.4.02

20. S. Samanta, P. Banerji, P. Ganguly, Focused ion beam fabrication of SU-8 waveguide structures on oxidized silicon. MRS Adv. **2**, 981–986 (2017)

21. S. Samanta, P.K. Dey, P. Banerji, P. Ganguly, Comparative Study between the results of effective index based matrix method and characterization of fabricated SU-8 waveguide. Opt. Commun. **382**, 632–638 (2017)

22. Standard Operating Procedure for Piranha Solutions, Univ. of Maryland. Available: http://www.lamp.umd.edu/Sop/Piranha_SOP.htm. Accessed on May 2016

23. Oxide Growth, Univ. of Florida. Available: http://www.che.ufl.edu/unit-ops-lab/experiments/semiconductors/oxide-growth/Oxide-growth-theory.pdf. Accessed on May 2016

24. Microchem website, http://www.microchem.com/. Accessed on Feb 2015

25. P.R. Babu, Infinite impulse response filters, in *Digital Signal Processing* (2011, Scitech Publication Pvt., Chennai, India), pp. 5.1–5.17

26. P. Ganguly, C.L. Sones, Y.J. Ying, H. Steigerwald, K. Buse, E. Soergel, R.W. Eason, S. Mailis, Determination of refractive indices from the mode profiles of UV-written channel waveguides in LiNbO3-crystals for optimization of writing conditions. J. Lightwave Technol. **27**, 3490–3497 (2009)

27. T. Ghosh, B. Samanta, P.C. Jana, P. Ganguly, Determination of refractive index profile and mode index from the measured mode profile of single-mode $LiNbO_3$–diffused waveguides. Fiber Integr. Opt. **31**, 1–10 (2012)

28. B. Yang, L. Yang, R. Hu, Z. Sheng, D. Dai, Fabrication and characterization of small optical ridge waveguides based on SU-8 polymer. J. Lightwave Technol. **27**, 4091–4096 (2009)

29. D. Dai, B. Yang, L. Yang, Z. Sheng, Design and fabrication of SU-8 polymer-based micro-racetrack resonators, in *Proceeding SPIE*, vol. 713 (The International Society for Optical Engineering, 2008), pp. 713414

30. T.A. Anhoj, J. Hubner, A.M. Jorgensen, Fabrication of high aspect ratio SU-8 structures for integrated spectrometers. Ph.D. Thesis, Technical University of Denmark (2007)

Chapter 4
Design and Development of Some SU-8 Wire Waveguide Structures

4.1 Introduction

This chapter deals with the design and development of wire waveguide structures, viz. directional coupler and micro-ring resonator using SU-8 polymer [1–3]. For optical integrated circuits based on micro-ring resonators (MRRs), one needs to know the coupling coefficients between straight and curved waveguides, and two curved waveguides accurately to compute the resonance characteristics of a single micro-ring, and two coupled micro-rings [4, 5]. For this application, the design and analysis of directional couplers consisting of two straight waveguides, straight and curved waveguides, and both curved waveguides (parabolically weighted coupling) are considered here, whereas the micro-ring resonator discussed in this chapter was made up of two bus waveguides with a ring waveguide in between. Fabrication of these structures was done using chrome mask by optical lithography, instead of laser direct writing technique, to achieve better precision and control over fabricated waveguide structures. In all waveguide devices presented in this chapter, we had used a plasma-enhanced chemical vapor deposited (PECVD) thick oxide buffer layer to reduce light leakage into silicon substrate. The minimum gap between the coupled air-cladded SU-8 waveguides achieved by optical lithography was around 0.57 μm. Some of the characterization results were validated with the simulated ones (for both TE and TM polarizations). The fabricated micro-ring resonator (MRR) was characterized using semiconductor laser diode and a monochromator. It was observed that MRR can be useful as a band-pass filter around 1565 nm wavelength of light with a 3-dB bandwidth of 5.36 nm for TE polarization. A feasibility study on photonic crystal structure fabricated on straight SU-8 wire waveguide was also performed both theoretically and experimentally. These may be useful as conventional photonic crystal waveguide or as input/output light coupler in an optical integrated circuit.

© The Author(s), under exclusive license to Springer Nature Singapore Pte Ltd. 2020
S. Samanta et al., *Photonic Waveguide Components on Silicon Substrate*,
SpringerBriefs in Applied Sciences and Technology,
https://doi.org/10.1007/978-981-15-1311-4_4

4.2 Optical Directional Couplers

Directional coupler is one of the important basic components of optical integrated circuits (OICs) which is composed of two evanescently coupled waveguide arms typically consisting of a straight interaction region between two bent transition regions [6, 7]. It is extensively used in all sorts of optical passive/active components, like optical power splitters, optical modulators, wavelength division multiplexers, add—drop multiplexers, micro-ring resonators, and Mach–Zehnder interferometers [4, 8–10]. Straight-type directional coupler consists of two adjacent parallel straight waveguides with a small separation between them. The amount of coupling between waveguides depends on fabrication parameters of the coupler, i.e., waveguide widths, gap between them, coupling length, and light wavelength and polarization.

4.2.1 Design

The key parameter for the design of a directional coupler consisting of two straight waveguides is its critical coupling length (L_c), which is the minimum length of coupling between the adjacent single-mode channel waveguides necessary for complete transfer of power to the coupled waveguide. To compute L_c, the lateral effective index profile of two straight coupled waveguides was computed and subsequently used in transfer matrix method [11] to get symmetric (β_s) and antisymmetric (β_a) propagation constant values.

In the coupling coefficient versus propagation constant plot, we obtained two Lorentzian peaks, one for each of them. Then critical coupling length of the directional coupler may be obtained as [12]:

$$L_c = \frac{\pi}{\beta_s - \beta_a} \tag{4.1}$$

Figure 4.1 shows coupled curved waveguides with minimum separation, S_0, at the middle and maximum separation, S_1, at the ends where coupling is very weak and negligibly small, g_0 is the input optical wave, and g_1 and g_2 are the output optical waves in waveguides 1 and 2, respectively, R is the radius of curvature of the waveguides, total length (L) of the bend waveguide is $2R\theta$, where 2θ is the angle made by the curved waveguides at the center of curvature. The radii of curvature for these waveguides are expected to be low enough due to the high contrast of the waveguides. The differential equations for co-directional coupling between waveguides are [14]:

$$dg_1/dy - j\delta g_1 = -jk(y)g_2 e^{-j\varnothing(y)} \tag{4.2}$$

$$dg_2/dy + j\delta g_2 = -jk(y)g_1 e^{-j\varnothing(y)} \tag{4.3}$$

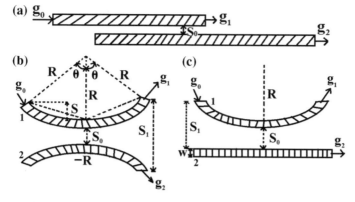

Fig. 4.1 Coupled waveguides: **a** two straight waveguides, **b** two curved waveguides, **c** one straight and one curved waveguide (reprinted with permission from [13]. ©2015 Elsevier GmbH)

where $\varnothing(y)$ is the phase and $k(y)$ is the coupling coefficient. To analyze the coupling between the curved waveguides, the two coupled Eqs. (4.2) and (4.3) were transformed into a single nonlinear Riccati equation [15, 16]:

$$d\rho/dy = -j(2\delta + d\phi/dy)\rho + jk(-1 + \rho^2) \tag{4.4}$$

where $\rho = (g_2/g_1)\exp^{-j\phi}$ and $\delta = (\beta_2 - \beta_1)/2$ are the detuning parameter, y is the distance along the propagation direction. The critical coupling length between two straight single-mode coupled waveguides was computed for different separations, and maximum separation S_1 was chosen such that there was practically no coupling between the waveguides. The maximum separation between the input and output ends of the directional coupler can be written as:

$$S_1 = S_0 + 2S = S_0 + 2R\sin^2(\theta/2) \tag{4.5}$$

$$S_1 = S_0 + S = S_0 + R\sin^2(\theta/2) \tag{4.6}$$

where Eq. (4.5) is for two curved waveguides and Eq. (4.6) is for one straight and one curved waveguides. For two similar waveguides, $\delta = 0$, and Eq. (4.4) yields:

$$|g_1|^2 = \cos^2(C) \tag{4.7}$$

$$|g_2|^2 = \sin^2(C) \tag{4.8}$$

where $|g_1|^2 + |g_2|^2 = |g_0|^2 = 1$ and $C = \int_{-L/2}^{L/2} \kappa(y)dy$ are the overall coupling coefficient; L being the total length of the curved waveguide directional coupler. Thus from Eq. (4.8), one may compute the normalized power coupling to the second waveguide. Equation (4.8) indicates that the crossover intensity is periodic with C, and by adjusting R and S_0, one may get the desired crossover intensity.

4.2.2 Computed Results

Figure 4.2 shows the computed data of critical coupling lengths (L_c) for different separations of directional couplers consisting of two straight waveguides for both TE and TM polarizations. Since TE modes are more confined (coupling between the waveguides is weaker) than TM modes, L_c is appreciably larger for TE modes. Typical resonance characteristics of two straight waveguide directional couplers computed by EIMM are shown in Fig. 4.3, where gap between waveguides in coupled region is 0.5, 0.6, and 0.7 μm. Computed results of coupling lengths by EIMM were validated with 2D-FDTD method. These computed coupling coefficients κ ($=\pi/2L_c$) were used to evaluate the overall coupling coefficient C of the curved waveguides directional couplers. The minimum and maximum separations between straight and curved waveguides were chosen as 0.5 μm and 1.0 μm, respectively, as beyond that coupling coefficient κ was found to be negligibly small.

The power distributions between the output ports of the straight and curved waveguide directional couplers and coupler with both curved waveguides for different radii of curvature are shown in Fig. 4.4 for TE mode. It is observed that 100% optical power transfer between the straight and curved waveguides occurred for a bending radius of 56 μm, while for both curved waveguides the value is 115 μm. It seems from our computations that for a fixed radius of curvature of waveguides, any amount of power coupling in the second waveguide may be achieved by adjusting the minimum gap S_0 between them. Similar analysis of the directional couplers may also been done for TM mode.

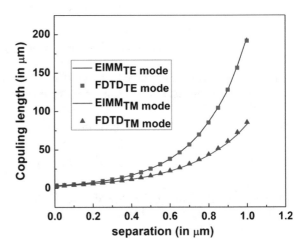

Fig. 4.2 Coupling length versus separation between two straight SU-8 waveguides for TE and TM mode ($w = 0.6\ \mu$m, $h = 1.5\ \mu$m)

Fig. 4.3 Typical resonance characteristics of two straight-waveguide directional couplers with separation between waveguides in coupled region **a** 0.5 μm, **b** 0.6 μm, **c** 0.7 μm

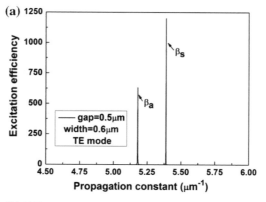

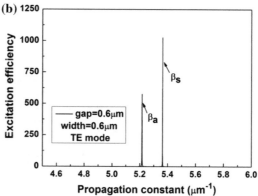

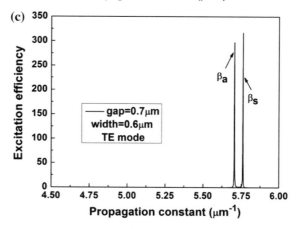

Fig. 4.4 Propagation
distance versus relative
optical power of output
optical wave g_2 for TE mode
for **a** one straight and one
curved, **b** two curved SU-8
waveguides

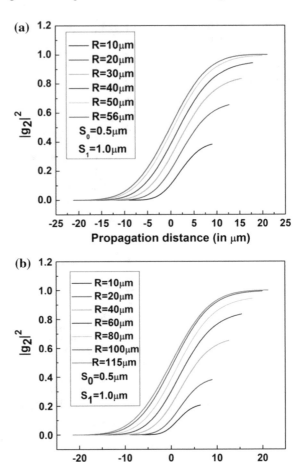

4.2.3 Fabrication

Figure 4.5 shows the flowchart of the fabrication process of SU-8 waveguide structure onto the wafer surface. From step-2 to step-6 fabrication, the process is same as given in Sect. 3.2.1. Chrome mask plate (3×3 inch) was used for patterning of wire optical waveguides and directional couplers of waveguide dimension around 1 μm. In order to write a dark-field mask of required pattern, an in-house laser writing system (Microtech laser writer, LW-2002) fitted with a He–Cd laser source emitting 405 nm wavelength of light was used. The exposure dose and writing speed were standardized to obtain the required dimensions of the dark-field mask of waveguide structures; the calibrated gain for this patterning was 5.3, and the corresponding bias was 98 mJ/cm^2. After completion of writing, the mask plates were carefully developed for 45 s in a developer solution (23:2 volume ratio of HPRD and DI water)

Fig. 4.5 Fabrication process flow using chrome mask (reprinted with permission from [3]. ©2018 Elsevier B.V.)

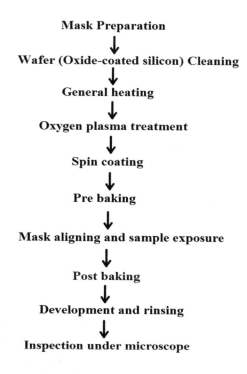

Mask Preparation

↓

Wafer (Oxide-coated silicon) Cleaning

↓

General heating

↓

Oxygen plasma treatment

↓

Spin coating

↓

Pre baking

↓

Mask aligning and sample exposure

↓

Post baking

↓

Development and rinsing

↓

Inspection under microscope

for 45 s to remove resist from exposed regions; and finally, the uncovered chromium was etched in chrome etchant for about 5 min to obtain the final pattern. Thereafter, it was thoroughly rinsed in DI water, and the protective positive resist on the mask plate was stripped by acetone at 70 °C. For all fabricated waveguides reported in this chapter, we had used a (100) silicon wafer with 3-μm-thick plasma-enhanced chemical vapor deposited (PECVD) oxide layer as our substrate. It was fabricated for us by Semi-Conductor Laboratory, Chandigarh, India. The oxide thickness of 3 μm was chosen in order to reduce the leakage of the guided mode from top SU-8 layer to silicon substrate. After cleaning and drying, the substrate was placed in a plasma cleaner (Zepto, Diener Electronic, Germany) for 1.5 min at 40 W power with 1.7 mbar oxygen gas pressure and 15 sccm gas flow rate to improve the adhesion of SU-8 with SiO$_2$ wafer. Next to SU-8 coating and prebaking, the sample was exposed to UV light (365 nm) for 5 s through the mask plate of the structure using a mask aligner (MA6, Karl Suss, Germany). After exposure, post-baking was done at 65 °C for 1 min and next 1 min at 95 °C. The wafer was then developed using SU-8 developer for 3 s and rinsed with isopropyl alcohol (IPA) to confirm complete development. During fabrication, careful optimization was made regarding exposure dose and time, and development time on which opening of the gap between waveguides depends. After thorough inspection under microscope, the fabricated waveguide structure on SiO$_2$ was cleaved from both edges for proper light coupling in the waveguide structures.

4.2.4   Characterization

The waveguide pattern was inspected with both Dektak surface profiler (Dektak 150, Veeco Sloan) and MERLIN ZEISS field-emission scanning electron microscope (FESEM). Before inspecting under the FESEM, the wafers were coated with a thin layer of gold–palladium (80:20%) alloy in Quorum gold–palladium coater for about 2 min, as SU-8 is a polymer which tends to charge when scanned by electron beam; also palladium enhances the ultimate resolution performance by restricting agglomeration of gold during deposition. The width of the SU-8 waveguide as found from FESEM was ~3.5 μm. The wafer was tilted by 20° to see the waveguide edge and to measure the waveguide thickness. From both the measurements using surface profiler and FESEM, the waveguide thickness was found around 2 μm. It may be mentioned that the thickness of waveguide depends also on SU-8 2005 aging, which deteriorates with time. The FESEM view of the waveguide edges is shown in Fig. 4.6. The photomicrograph of a fabricated two-waveguide directional coupler of 100 μm coupling length and 0.57 μm separation between the waveguides in the coupled region is shown in Fig. 4.7.

Fig. 4.6 FESEM view of SU-8 waveguide edge **a** top view, **b** tilted at 20° (reprinted with permission from [3]. ©2018 Elsevier B.V.)

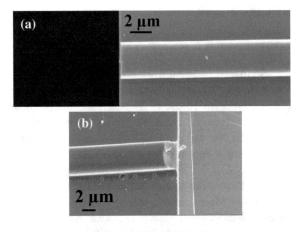

Fig. 4.7 FESEM view of the coupled region of the directional coupler

Fig. 4.8 **a** Mode image of single-mode SU-8 waveguide, **b** mode image with diffraction pattern

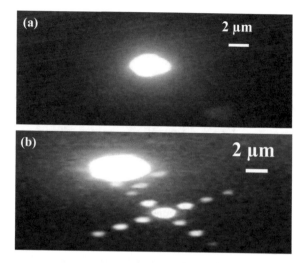

4.2.4.1 Mode Output Observation

A semiconductor laser source of 1.55 μm was coupled to the waveguide structure through single-mode optical fiber using a high-resolution micropositioner; and the output was observed in a monitor through 20× objective and infrared camera. The single-mode waveguide output is shown in Fig. 4.8a. We observed a diffraction pattern as shown in Fig. 4.8b, in some cases during our experimentation. The reason for the occurrence of this diffraction pattern was due to the fact that the operating wavelength and dimensions of waveguide were of the same order (repeated experiments were done for its verification). However, it may also occur for other optical component in the ray path.

4.2.4.2 Coupling Length Measurement

For the coupling length measurement, we had taken a straight wire SU-8 waveguide as the reference and assumed the input of the directional coupler to be equal to the output of reference straight waveguide. Now, when the separation between the adjacent parallel waveguides is small, light will couple into the second waveguide. However, during the coupling process, it may happen that all the power is not transferred at output of directional coupler; some portion of light remains in the coupling region. So, this excess loss is the difference between the directional coupler output and input and this can be written as Excess loss $= -10 \log_{10} (P_o/P_i)$; where P_o is the output power and P_i is the input of the coupler. Now, $P_o/P_i = -10^{\text{Excess loss}/10} = \sin^2 (\kappa L)$; κ being the coupling coefficient, and L is the physical length (100 μm as designed in mask and replicated in wafer). As we know, $\kappa = \pi/2L_c$, so coupling length can be measured by this relation. The abruptly ending waveguides emit light at the end of the waveguide which is not likely to perturb the other waveguide mode.

Table 4.1 Comparison of characterized results of SU-8 wire waveguide fabricated by optical lithography with other authors' work

Waveguide dimension (μm)		Top clad	Operating wavelength (nm)	Propagation loss (dB/mm)	Reference
Thickness	Width				
4.5	5.0	Mr-L 6050XP	800–1600	0.02–0.3	[17]
3.3	3.3	Polysiloxane	1550	0.155	[18]
2.0	4.0	NOA-61	1550	0.125	[19]
1.7	2.8	Air	1550	0.1	[20]
2.0	3.5	Air	1550	0.8	Our work [1]

The waveguide propagation loss of single-mode SU-8 wire waveguide for both TE and TM modes was measured by cutback method using chopper–detector–lock-in amplifier arrangement (same setup is shown in Fig. 3.2); the respective values in TE and TM polarization were found to be 0.84 and 0.77 dB/mm. The measured coupling length of directional coupler with 0.6 μm separation was found to be 40 μm and 31 μm for TE and TM mode, respectively, while its 300 μm (TE mode) and 65 μm (TM mode) for 1 μm separation between the coupled waveguides. We had compared these characterized results with other published works (as shown in Table 4.1), which showed that our measured loss is a bit higher. However, it was comparable with the other works keeping in mind that optical loss depends on the waveguide dimensions and top clad layer, processing parameters of SU-8 and even aging effect of SU-8.

4.2.4.3 Comparison Between Fabricated and Theoretical Results

A comparison was made between the measured data with the simulated ones. Table 4.2 shows the detailed comparison for both TE and TM mode with 0.6 and 1 μm separations. We can see from the table that the experimental results match reasonably well with the theoretical values as obtained using effective index-based matrix method (EIMM) for smaller separation of 0.6 μm. For 1 μm separation, the difference between the experimental and theoretical results is larger. It is expected due to the fact that coupling length critically depends on gap between the waveguides.

Table 4.2 Coupling lengths for different separations

Separation (μm)	Coupling length (μm)			
	TE mode		TM mode	
	Experimental value	Theoretical value by EIMM	Experimental value	Theoretical value by EIMM
0.6	40	37.76	31	21.97
1.0	300	195.23	65	82.87

As gap increases, L_c increases; so more variation was observed between experimental and simulated values for increased gap between waveguides.

4.2.4.4 Discussions

A study on design, fabrication, and characterization of straight-waveguide directional coupler was done using SU-8 polymer waveguides, which operated in single-mode region at 1.55 μm transmitting wavelength. The measured total insertion losses of the fabricated waveguides were fairly low. The measured coupling length of fabricated directional coupler matched reasonably well with our theoretical prediction using effective index-based matrix method. Using these results, optical micro-ring resonator was developed, which is elaborated in the following section.

4.3 Micro-ring Resonator

Waveguide device like micro-ring resonator (MRR) is basically a ring waveguide acting as the resonant cavity with one or two bus waveguides acting as input and output ports. The coupling mechanism involved in this device is the evanescent coupling between ring and adjacent bus waveguide [21]. This may be vertically coupled or laterally coupled; both configurations have certain pros and cons. In case of laterally coupled resonator, both the ring and bus waveguides lay on the same horizontal plane and thus require very accurate lithography and etching processes in order to obtain submicron gap between bus and ring, thereby limiting the flexibility in the device design. On the other hand, ring and bus waveguides in vertical configuration do not lie in the same plane. Ring is placed on top or bottom of bus waveguides; as a result, the ring and bus may be of different material, and the thickness need not be the same and can be controlled accurately during deposition—all these enhance the design freedom. However, this vertical configuration is expensive due to the additional processing step of the ring in contrast to lateral configuration which requires only a single layer. Moreover, fabrication of vertically coupled ring resonator is complex as wafer bonding and regrowth are required to manufacture these devices; also alignment is an issue as there are two processing steps [22, 23]. In most of the previously reported research articles [24–26], SU-8 waveguide-based MRRs were fabricated by costly electron beam lithography and characterized by a very narrow bandwidth tunable laser source (TLS) and photodiode (PD) or optical spectrum analyzer (OSA). Here we made an attempt to develop SU-8 wire waveguide-based MRR by 365 nm I-line optical lithography, and characterization was performed by using semiconductor laser source, monochromator, and InGaAs detector. Design part of this device was performed by using effective index-based matrix method (EIMM) [13, 27] and coupled mode theory (CMT) [28, 29]. We had used a silicon dioxide (SiO_2) deposited silicon (Si) wafer as our substrate material.

The design, fabrication, and characterization of the horizontal configuration MRR are discussed in the following sections.

4.3.1 Design

The schematic of the micro-ring resonator is shown in Fig. 4.9; 1 denotes the input port, 2 is the through port, 3 and 4 are the drop and add ports, respectively. Multiple wavelengths $\lambda_1, \lambda_2, \ldots, \lambda_0, \ldots, \lambda_n$ input into terminal 1. The wavelength which meet the resonant condition, i.e., $n_g\, l = m\, \lambda_0$ (n_g is the group index, $l = 2\pi R$ is the circumference, R being the radius of curvature, m is the mode number), will couple with the ring, and others will pass through the terminal 2. So, if λ_0 meets the resonant condition, the coupling of wave with wavelength λ_0 will be enhanced; and all other wavelengths will be suppressed; thus only λ_0 will be dropped into port 3.

The micro-ring device was designed using the following steps: (i) waveguide design using effective index-based matrix method (EIMM) [13, 27]; (ii) determination of coupling coefficient between straight and curved waveguides [13]; (iii) determination of bending losses of bent SU-8 wire for different radii of curvature by conformal mapping [30] and EIMM.

The through port and drop port power transmission responses (T_{through} and T_{drop}, respectively) were calculated from the following relations [31]:

$$T_{\text{through}} = \frac{(\lambda - \lambda_0)^2 + \left(\frac{\text{FSR}}{4\pi}\right)^2 \left(\kappa_p^2\right)^2}{(\lambda - \lambda_0)^2 + \left(\frac{\text{FSR}}{4\pi}\right)^2 \left(2\kappa^2 + \kappa_p^2\right)^2};$$

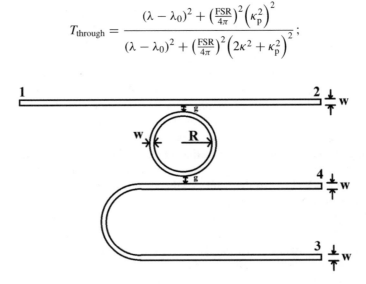

Fig. 4.9 Schematic of laterally coupled micro-ring resonator with through and drop ports (reprinted with permission from [3]. ©2018 Elsevier B.V.)

$$T_{\text{drop}} = \frac{4\left(\frac{\text{FSR}}{4\pi}\right)^2\left(\kappa^4\right)}{(\lambda - \lambda_0)^2 + \left(\frac{\text{FSR}}{4\pi}\right)^2\left(2\kappa^2 + \kappa_{\text{p}}^2\right)^2} \tag{4.9}$$

These relations are valid for wavelengths close to λ_0, the center resonance wavelength. *FSR* is the free spectral range which is defined as the distance between the adjacent resonant peaks, κ^2 is the power coupling coefficient between the bus and ring waveguides, and κ_{p}^2 is the propagation power loss coefficient per round trip in the ring resonator which includes propagation loss and bending loss of the ring. Free spectral range (FSR), quality factor (Q), and extinction ratio (ER) of the resonator can be expressed as [32]:

$$\text{FSR} = \frac{\lambda_0^2}{n_g L}; \ Q = \frac{\lambda_0}{\text{FWHM}} ; \ \text{ER} = -10\log_{10}\left[1 - \left\{\left(\frac{\sqrt{1-\kappa^2}-\kappa_p^2}{\sqrt{1-\kappa^2}+\kappa_p^2}\right)\left(\frac{1+\kappa_p^2\sqrt{1-\kappa^2}}{1-\kappa_p^2\sqrt{1-\kappa^2}}\right)\right\}^2\right] \tag{4.10}$$

Here FWHM is the full-width-at-half-maxima at 3-dB point, and n_g represents the group index in SU-8 waveguides. The width (w) of the bus and ring waveguides was taken such that the device supports single-mode operation at 1.55 µm transmitting wavelength, g is the gap between the bus and ring, and the radius (R) was chosen such that the bending loss becomes negligible.

4.3.2 Computed Results

The radius of curvature of the ring that we had considered was 15 µm as the bending loss at this radius was negligibly small (as shown in Fig. 2.15 of this book). Considering the fabrication tolerances, the separation between ring and bus waveguide was chosen 0.5 µm, and the waveguide width was 3.5 µm. The computed power coupling coefficient between ring and bus waveguide at the chosen radius was 0.541607, and the propagation power loss coefficient per round trip in the ring was 0.0134767. This power loss coefficient includes both bending loss and propagation loss. The calculated bending loss at 15 µm radius was 10^{-4} dB/µm, while propagation loss value was taken from our obtained minimum experimental data, which was 0.5 dB/mm (as described in Chap. 3). Figure 4.10 depicts the transmission characteristics for the ring resonator as computed using Eqs. (4.8) and (4.9). The computed through port notch of the structure was –87.98 dB at a resonating wavelength of 1550 nm; and the computed free spectral range (FSR), quality factor (Q), and extinction ratio (ER) of the designed resonator were 16.79 nm, 30,312, and 13.745 dB, respectively, for transverse electric (TE) mode. During the design, we had taken n_g as equal to the effective refractive index of the straight waveguide, which was approximately same to the group index of the bent waveguide of MRR for 15 µm radius.

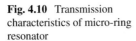

Fig. 4.10 Transmission
characteristics of micro-ring
resonator

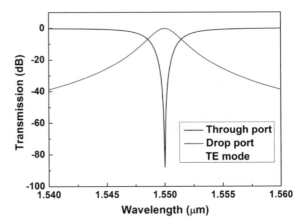

The resonance dip of the device depends on various factors, viz. temperature, physical deformation or compositional changes in waveguide core or cladding, which may result in shift of wavelength.

The temperature sensitivity of the micro-ring resonator is also studied here. Figure 4.11 shows the resonance shift of spectrum with variation of temperature. More the temperature deviation (δT) from room temperature (300 K), more the spectrum shifts toward left. This blueshift is obvious due to the negative thermo-optic coefficient of SU-8 polymer [33]. Also, through port intensity varies with the change in temperature, as can be seen from Fig. 4.11. Since with the change in temperature there is a change in refractive index, and hence the resonance wavelength of the device, it can be used as a temperature sensor within a limited temperature range.

Fig. 4.11 Transmission
characteristics due to change
in temperature

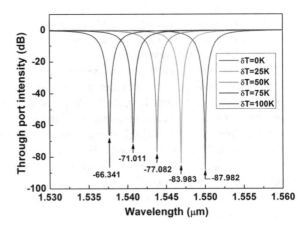

4.3.3   Fabrication

Fabrication process of the MRR was similar as stated in Sect. 4.2.3. To fabricate mask of the structure, data file was generated using AutoCAD software in CIF format which was subsequently transferred to LDF format to write a dark-field chromium mask. The microscopic view of fabricated structure is shown in Fig. 4.12a, b. It may be noted that before the fabrication of MRR by using chromium mask, we had also tried to fabricate the structure by laser direct writing at 375 nm writing wavelength. But since the laser spot size for direct writing was ~5 μm, it was not possible to fabricate a well-defined MRR. While inspecting in microscope, we observed prominent bulging in between the bus waveguide and the ring (Fig. 4.12c), which was due to the proximity error, i.e., due to the individual pattern features do not image independently, rather they interact with neighbor pattern features.

4.3.4   Characterization

The fabricated waveguide samples were manually cleaved by applying pressure with a sharp tweezer at 90° angle to the waveguides to obtain clean defect-free end-facets, which were needed to couple light into the device efficiently. To inspect different areas and for measurement of dimensions of the fabricated structure, an optical microscope (SG-V 84149, Union, Japan) and field-emission scanning electron microscope (FESEM) (Zeiss, Germany) were used. Figure 4.12a depicts the optical microscopic view of the fabricated micro-ring resonator. The obtained waveguide width and thickness were 3.5 μm and ~2 μm, respectively; minimum separation between the bus and ring was ~0.5 μm, the radius of the ring is being 15 μm. To obtain 0.5 μm minimum separation, the mask pattern and photolithographic process were critically calibrated. Figure 4.12b shows the bending region of 25 μm radius at the drop port of micro-ring resonator. It may be noted from Fig. 4.12a that in the coupling regions of micro-ring and bus waveguides, the waveguide widths are slightly more. This attributes to slight proximity effect still remained during exposure through mask pattern. Figure 4.12c was a preliminary attempt to fabricate the device using maskless laser direct writing technique with 375 nm UV laser source.

To get the near-field image of the device, an indium phosphide (InP)-based laser source (emitting ~1.5 mW continuous-wave power around 1.55 μm wavelength of light of linewidth ~20 nm) from CVI Melles Griot, USA, was coupled into input of the micro-ring resonator through a pigtailed single-mode optical fiber using a high precision three-axis micropositioner (ULTRAlign, Model: 561D, Newport, USA) and gimbal mount. After careful alignment of optical fiber output with the waveguide input, the outputs (through and drop ports) of the chip were imaged onto an IR camera (Model: 7290A, Electrophysics Micronviewer, USA) by a 20× objective lens. Figure 4.13 shows the recorded output image of the structure in a monitor (ULTRAK) connected to the IR camera. Obtained two spots indicate through port

Fig. 4.12 Optical microscopic view of fabricated micro-ring **a** by optical lithography, **b** bend portion of drop port, **c** using laser direct writing technique (reprinted with permission from [3]. ©2018 Elsevier B.V.)

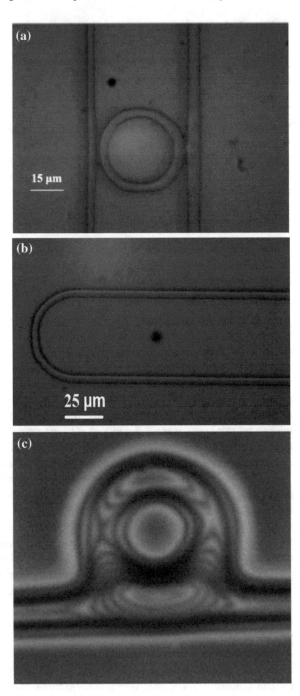

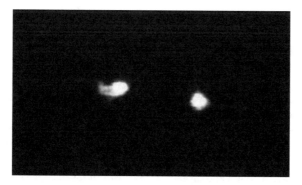

Fig. 4.13 Near-field image of the fabricated micro-ring resonator (reprinted with permission ©2018 Elsevier B.V.)

(left) and drop port (right) of the fabricated micro-ring resonator. Some wavelengths of light within laser emission spectra were available in drop port, whereas the rest of the wavelengths were in through port.

The through and drop port response of the chip were measured separately using an optical setup shown in Fig. 4.14. The outputs of each port were scanned by using a grating monochromator (Model No. 77200, Oriel Instruments, USA) and an InGaAs detector. A chopper (Model: SR540, Stanford Research Systems, USA)–lock-in amplifier (Model: SR 830 DSP, Stanford Research Systems, USA) arrangement was used to improve the signal-to-noise ratio (SNR) of measured data. The chopping frequency was regulated by chopper controller; and throughout our measurement process, it was fixed at 270 Hz. In our optical measurement setup, a Glan–Thompson polarizer was used to select either TE or TM mode of waveguide output. In addition, a calibration trace was done to eliminate the variations of laser power at different wavelengths and to overcome the wavelength dependency of optics and detector. During calibration, MRR chip was removed from the setup and fiber output (P_{in}) was directly scanned by monochromator. All the measurements were performed on a vibration-isolated optical breadboard at ~25 °C temperature.

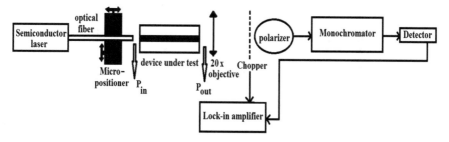

Fig. 4.14 Setup for measuring resonator response (reprinted with permission from [3]. ©2018 Elsevier B.V.)

Before taking any measurement, the monochromator itself was carefully calibrated by using a He–Ne laser output (0.6328 µm wavelength of light) and its harmonics. The resolution or bandpass of this monochromator was ~ 0.6 nm with minimum wavelength dial reading of 0.04 nm. Figure 4.15 shows the plot of measured through port response (P_{th}/P_{in}) of our fabricated micro-ring resonator for TM polarization. From this figure, two notches can be seen; one at 1544 nm and another at 1563 nm. The distance between the adjacent notches, i.e., free spectral range (FSR) is found to be ~19 nm (designed value: 16.34 nm), and the notches are obtained due to the resonating characteristics. Total insertion loss of the device for this TM mode was measured as 12.87 dB around 1.55 µm wavelength of light.

The normalized through and drop port responses of fabricated MRR for TE mode are shown in Fig. 4.16. The obtained FSR for this mode was ~16 nm (designed value: 16.76 nm). Since TE modes are in general more confined, in most of the applications MRR is used in TE polarization [23, 34, 35]. For TE mode, measured total insertion loss and extinction ratio of our fabricated device were 8.6 dB and 10.5 dB (designed value: 13.75 dB), respectively.

From Fig. 4.16, it may be noted that through port of this MRR can be used as an optical bandpass filter around 1565 nm transmitting wavelength with a 3-dB bandwidth of 5.36 nm. From our experiments, it has been observed that spectral response of through and drop ports of MRR is highly polarization-dependent, and the obtained Q value is ~292 for TE mode, which is far too low as compared to the expected value. There may be two reasons behind it. First, our SU-8 waveguides were having lateral widths of ~3.5 µm, which was not strictly a single-mode waveguide; although propagation losses of higher-order modes are more than the fundamental one. Second, the bandpass of our monochromator was quite high (~0.6 nm). From Fig. 4.16, it may be observed that obtained through port response is flat-top, having

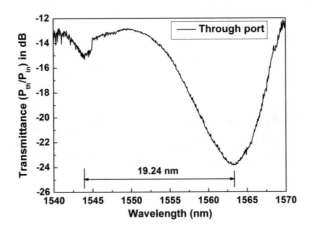

Fig. 4.15 Transmission characteristics for the fabricated micro-ring resonator for TM mode (reprinted with permission from [3]. ©2018 Elsevier B.V.)

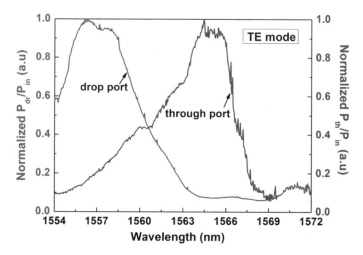

Fig. 4.16 Normalized through port and drop port response for TE mode (reprinted with permission from [3]. ©2018 Elsevier B.V.)

a width of ~1.5 nm. Hence, for more accurate measurement, such as with a high-resolution optical spectrum analyzer, one may obtain lower bandwidth and higher Q value of the same fabricated device. Also, obtained spectral responses of the through and drop port in Fig. 4.16 are pretty wide. The obtained Q value (~292) is too high for the spectral response of the waveguide structure consisting of directional couplers [36] and too low for micro-ring resonators, whereas measured FSR (16 nm) is also quite low compared to directional coupler structures. This may be due to the combined effect of optical mode of MRR, as well as the wavelength dependency of the waveguide structure. However, Q factor of the produced resonator is not that crucial for its application as a band-pass filter. A comparison regarding our characterized results was made with already published MRR papers; Table 4.3 shows the details of the comparison.

4.3.5 Discussions

Design, fabrication, and characterization of laterally-coupled circular micro-ring resonator were studied using SU-8 wire waveguides for TE/TM polarization. Design of MRR was carried out by using EIMM and coupled mode theory. Fabrication of the device was done by optical lithography using chrome mask and was characterized by using a grating monochromator and a semiconductor laser diode. Measured FSR and extinction ratio of MRR were reasonably close to the designed values, although obtained Q value was quite low as compared to an expected value. The waveguide width and thickness used in this device were 3.5 and 2.0 μm, and total insertion loss of the device both for through and drop port was measured as ~8.6 dB for TE

Table 4.3 Comparison of characterized results with other SU-8 MRR papers for TE polarization at 1.55 μm wavelength

MRR configuration	Process	Ring radius (μm)	Gap (nm)	Extinction ratio (dB)	FSR (nm)	Q factor	Insertion loss (dB)
Free-standing single-bus circular ring [24]	EBL and soft lithography	100	250	~9.0	2.435	2000	–
Single-bus circular ring [26]	EBL	200	350	~20.0	1.2	3555	9.9
Racetrack [33]	UV lithography	150	1000	~6.0	~1.0	8000	~5.0
Double-bus circular ring [3] (Our work)	UV lithography	15	500	10.5	16.0	292	8.6

polarization. The fabricated MRR can be useful as an optical bandpass filter. The measured 3-dB bandwidth of the filter output with a 0.6 nm bandpass monochromator was 5.36 nm around 1565 nm wavelength of light.

4.4 Photonic Crystal Structure on Waveguide

Photonic crystals are periodic structures with high dielectric and low dielectric regions, thus there is a periodic modulation of refractive index; the period being comparable to the wavelength of light in the material. One of the most interesting properties of photonic crystals is the capability of forbidding a certain frequency range of light (i.e., bandgap) from transmission [37]. When light falls in the bandgap region, it is unable to propagate in the crystal; while light on the surface is reflected, thus making it flexible for guiding light [38, 39]. The advent of these photonic bandgap materials not only enables molding of light flow, but also controls the dynamics of photonics; this allows manipulating the behavior of light. Photonic crystal finds its applications in optical processing and photonic integrated circuits [40, 41]. In this section, a feasibility study has been made to design and fabricate photonic crystal structure on SU-8 wire waveguide.

Fig. 4.17 Schematic of photonic crystal structure on waveguide

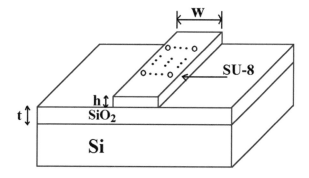

4.4.1 Design and Simulation

The two-dimensional (2-D) photonic crystal structure on SU-8 wire waveguide was designed and realized by using commercially available OptiFDTD software. Figure 4.17 shows the schematic of the designed structure; where w is the core width, t and h are being the respective oxide and core thickness. For this case, light is guided into the waveguide by total internal reflection due to the contrast in refractive indices of air (1.0)/SiO$_2$ (1.447) clads and core SU-8 (1.574) [42].

Starting with waveguide layout designer of OptiFDTD, profile and material, and the waveguide properties were defined; the waveguide width is being 4.5 μm. In the photonic bandgap crystal structure, 2-D rectangular lattice properties were chosen where the number of rows was taken as 6 and columns as 195. Following this, the atom properties were defined where elliptic waveguide was chosen and the radius was set at 250 nm. Then, with lattice constant of 800 nm and inserting vertical input plane (the wavelength being 1.55 μm), 2-D simulation with 12 lac time steps for both TE and TM modes resulted in the following transmission characteristics (Fig. 4.18a, b).

Next, with a slight change in the design parameters (i.e. by taking rows and columns as 5 and 100 respectively), polarization-independent bandgaps were obtained, where light won't transmit, rather light would reflect. Figure 4.19 shows the plane wave expansion (PWE) band solver result of the same. It can be seen from this figure that the wavelength ranging from 0.579 to 0.703 μm and from 1.692 to 2.277 μm acts as the band stop region irrespective of any polarization.

4.4.2 Fabrication

Fabrication of optical waveguide with photonic crystal structure was carried out using optical lithography followed by focused ion beam (FIB) lithography. First of all, SU-8 waveguide was fabricated by photolithography using chrome mask (as described in Sect. 4.2.3). Next, on the top of this waveguide, photonic crystal structure was

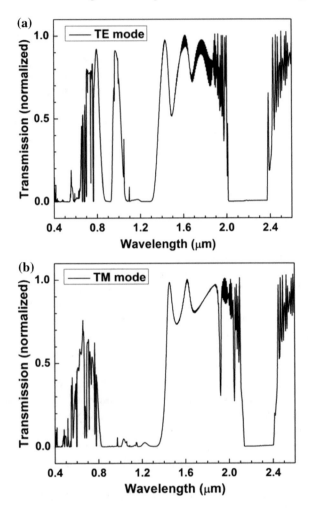

Fig. 4.18 Transmission characteristics **a** TE mode, **b** TM mode

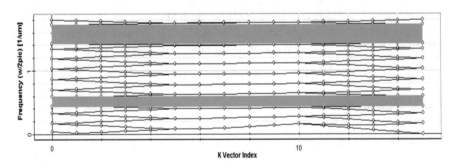

Fig. 4.19 Polarization-independent band gaps using PWE band solver

written directly by focusing Ga$^+$ ions using AURIGA Compact Zeiss focused ion beam system with an acceleration voltage of 30 kV, and beam current and ion dose used were 200 pA and 10,000 C/cm^2, respectively.

4.4.3 Characterization

Figure 4.20 is the scanning electron microscopic (SEM) view of focused ion beam spot size, which was taken as ~160 nm. Figure 4.21 is the fabricated photonic crystal on SU-8 waveguide, while Fig. 4.22 shows the atomic force microscopic (AFM) view of 2-D profile of the photonic crystal. A milling depth of ~220 nm was obtained (single-pass milling strategy was used), though AFM measurement of milling depth has some uncertainties. According to Benisty et al. [43], the depth of the photonic crystal holes should be large enough to overlap completely with the vertical profile of the guided mode so that extrinsic losses can be minimized. So the required depth of the holes will be around 1.5 μm in the present study. The drilled holes have non-circular

Fig. 4.20 Spot size of ion beam

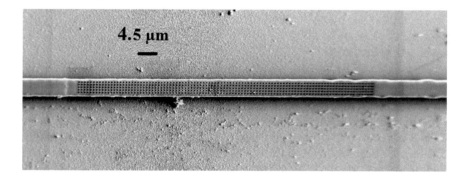

Fig. 4.21 Fabricated photonic crystal structure on SU-8 wire waveguide

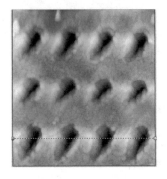

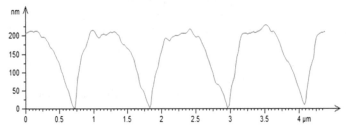

Fig. 4.22 AFM view of 2-D profile of the fabricated photonic crystal

(conical) shape; this is due to the re-deposition effect during milling and reflection of Gallium ions from the sidewall. This can be overcome if the holes are etched into the silicon dioxide layer. The guided light passing under the holes is avoided, thus results in circular (rather cylindrical) holes [44]. Re-deposition of material can also be reduced by using multipass milling technique and helps in obtaining cylindrical holes [45]. To minimize extrinsic losses (include the extra out-of-plane scattering resulting from a non-ideal hole shape), it is essential that the depth of the photonic crystal holes is large enough to overlap completely with the vertical profile of the guided mode [43]. Each hole of the photonic crystal waveguide was of ~460 nm in diameter, the periodicity was ~970 nm, and the rms roughness being 33.1 nm.

4.4.4 Discussions

Focused ion beam lithography is one of the promising techniques for fabricating nanometer-level features like photonic crystals. We made use of this technique along with optical lithography to fabricate our designed photonic crystal structure on top of SU-8 wire waveguide. The aspect ratio of the periodic structure was chosen such that the wall between holes did not get collapsed; also taking into account, the bandgap was not too narrow. This structure can be used as a polarization-independent optical band-pass filter for wavelength ranging from 0.703 to 1.692 μm. This kind of photonic crystal structure may also be used to couple and/or decouple light into a wire waveguide [46, 47], instead of using end-fire coupling with optical fiber or microscope objective, which require critical end-facet polishing steps. Although detail optical characterization of this photonic crystal waveguide was not performed in this book, the fabrication effort certainly confirms its feasibility [48].

4.5 Conclusions

The SU-8 wire waveguide-based air-cladded directional coupler and micro-ring resonator (lateral configuration) were designed, analyzed, fabricated, and characterized. Fabrication was done by optical lithography using a patterned chrome mask to reduce the waveguide widths and to achieve submicron separation between coupled waveguides. A 3-μm-thick plasma-enhanced chemical vapor-deposited (PECVD) silicon dioxide buffer layer on a silicon wafer was used to reduce light leakage into silicon wafer. The fabricated MRR was characterized by using a semiconductor laser diode and a simple monochromator of ~0.6 nm resolution. The characterization result indicated that this MRR can be useful as an optical band-pass filter. Some theoretical studies to fabricate photonic crystal structures on SU-8 wire waveguide were also conducted, and accordingly, a fabrication attempt of the structure was also made using FIB lithography.

References

1. S. Samanta, P.K. Dey, P. Banerji, P. Ganguly, Fabrication of directional coupler using SU-8 wire waveguide by optical lithography, in *International Conference on Fiber Optics and Photonics* (IIT Kanpur, India, 2016), p. W3A.87
2. S. Samanta, P. Banerji, P. Ganguly, Micro-ring resonator using SU-8 waveguides for temperature sensor, in *International Conference on Fiber Optics and Photonics* (IIT Kanpur, India, 2016), p. W2F.4
3. S. Samanta, P.K. Dey, P. Banerji, P. Ganguly, Development of micro-ring resonator-based optical bandpass filter using SU-8 polymer and optical lithography. Opt. Mater. **77**, 122–126 (2018)
4. B.E. Little, S.T. Chu, H.A. Haus, J. Foresi, J.-P. Laine, Microring resonator channel dropping filters. J. Lightwave Technol. **15**, 998–1005 (1997)
5. A. Delage, D.-X. Xu, R.W. McKinnon, E. Post, P. Waldron, J. Lapointe, C. Storey, A. Densmore, S. Janz, B. Lamontagne, P. Cheben, J.H. Schmid, Wavelength-dependent model of a ring resonator sensor excited by a directional coupler. J. Lightwave Technol. **27**, 1172–1180 (2009)
6. J.P. George, N. Dasgupta, B.K. Das, Compact integrated optical directional coupler with large cross section silicon waveguides, in *Silicon Photonics and Photonic Integrated Circuits II, Proceedings of SPIE Photonics Europe*, vol 7719 (2010), p. 77191X
7. W.J. Chen, S.M. Eaton, H. Zhang, P.R. Herman, Broadband directional couplers fabricated in bulk glass with high repetition rate femtosecond laser pulses. Opt. Exp. **16**, 11470–11480 (2008)
8. K. Kubota, J. Noda, O. Mikami, Traveling wave optical modulator using a directional coupler LiNb0$_3$ waveguide. IEEE J. Quant. Electron. **QE-16**, 754–760 (1980)

9. R.J. McCosker, G.E. Town, WDM for fluorescence biosensing using a multi-channel directional coupler, in *Optical Sensors* (2010), p. SWB4
10. J. Wang, L.R. Chen, Low crosstalk Bragg grating/Mach-Zehnder interferometer optical add-drop multiplexer in silicon photonics. Opt. Express **23**, 26450 (2015)
11. A. Ghatak, K. Thyagarajan, M.R. Shenoy, Numerical analysis of planar optical waveguides using matrix approach. J. Lightwave Technol. **5**, 660–667 (1987)
12. H. Nishihara, M. Haruna, T. Suhara, *Optical in Tegrated Circuits* (McGraw-Hill, New York, 1987)
13. S. Samanta, P. Banerji, P. Ganguly, Effective index-based matrix method for silicon waveguides in SOI platform. Optik Int. J. Light Electron Opt. **126**, 5488–5495 (2015)
14. G. Lifante, *Integrated Photonics: Fundamentals* (Wiley, England, 2003)
15. H. Kogelnik, Filter response of nonuniform almost-periodic structures. Bell Syst. Tech. J. **55**, 109–126 (1976)
16. R.C. Alferness, P.S. Cross, Filter characteristics of codirectionally coupled waveguides with weighted coupling. IEEE J. Quantum Electron. **14**, 843–847 (1978)
17. M. Nordstrom, D.A. Zauner, A. Boisen, J. Hubner, Single-mode waveguides With SU-8 polymer core and cladding for MOEMS applications. J. Lightwave Technol. **25**, 1284–1289 (2007)
18. S. Madden, Z. Jin, D. Choi, S. Debbarma, D. Bulla, B. Luther-Davies, Low loss coupling to sub-micron thick rib and nanowire waveguides by vertical tapering. Opt. Express **21**, 3582–3594 (2013)
19. K.K. Tung, W.H. Wong, E.Y.B. Pun, Polymeric optical waveguides using direct ultraviolet photolithography process. Appl. Phys. A Mater. Sci. Process. **80**, 621–626 (2005)
20. B. Yang, L. Yang, R. Hu, Z. Sheng, D. Dai, Fabrication and characterization of small optical ridge waveguides based on SU-8 polymer. J. Lightwv. Technol. **27**, 4091–4096 (2009)
21. C.Y. Chao, W. Fung, L.J. Guo, Polymer microring resonators for biochemical sensing applications. J. Sel. Top. Quantum Electron. **12**, 134–142 (2006)
22. O.G. Lopez, D.V. Thourhout, D. Lasaosa, M. Lopez-Amo, R. Baets, M. Galarza, Vertically coupled microring resonators using one epitaxial growth step and single-side lithography. Opt. Exp. **23**, 5317–5326 (2015)
23. M. Balakrishnan, E.J. Klein, M.B.J. Diemeer, A. Driessen, Fabrication of an electro-optic polymer microring resonator, in *Proceedings of Symposium IEEE/LEOS Benelux Chapter* (2006), pp. 73–76
24. Y. Huang, G.T. Paloczi, A. Yariv, C. Zhang, L.R. Dalton, Fabrication and replication of polymer integrated optical devices using electron-beam lithography and soft lithography. J. Phys. Chem. B **108**, 8606–8613 (2004)
25. G.T. Paloczi, Y. Huang, A. Yariv, Free-standing all-polymer microring resonator optical filter. Electron. Lett. **39**, 1650–1651 (2003)
26. J.K.S. Poon, Y. Huang, G.T. Paloczi, A. Yariv, Soft lithography replica molding of critically coupled polymer microring resonators. IEEE Photon. Tech. Lett. **16**, 2496–2498 (2004)
27. P. Ganguly, Semi-analytical analysis of lithium niobate photonic wires. Opt. Commun. **285**, 4347–4352 (2012)
28. A. Yariv, Coupled-mode theory for guided-wave optics. IEEE J. Quantum Electron. **9**, 919–933 (1973)
29. H.A. Haus, W.P. Huang, S. Kawakami, N.A. Whitaker, Coupled-mode theory of optical waveguides. J. Lightwave Technol. **LT-5**, 16–23 (1987)
30. M. Heiblum, Analysis of curved optical waveguides by conformal transformation. IEEE J. Quantum Electron. **11**, 75–83 (1975)
31. S. Xiao, M.H. Khan, H. Shen, M. Qi, Modeling and measurements of losses in silicon-on-insulator resonators and bends. Opt. Exp. **15**, 10553–10561 (2007)
32. J. Niehusmann, A. Vorckel, P.H. Bolivar, T. Wahlbrink, W. Henschel, H. Kurz, Ultrahigh-quality-factor silicon-on-insulator microring resonator. Opt. Lett. **29**, 2861–2863 (2004)
33. D. Dai, B. Yang, L. Yang, Z. Sheng, Design and fabrication of SU-8 polymer-based micro-racetrack resonators, in *Proceedings of SPIE—The International Society for Optical Engineering*, vol. 7134 (2008), p. 713414

34. C. Delezoide, M. Salsac, J. Lautru, H. Leh, C. Nogues, J. Zyss, M. Buckle, I.L. Rak, C.T. Nguyen, Vertically coupled polymer microracetrack resonators for label-free biochemical sensors. IEEE Photonics Technol. Lett. **24**, 270–272 (2012)

35. T. Cai, Q. Liu, Y. Shi, P. Chen, S. Heb, An efficiently tunable microring resonator using a liquid crystal-cladded polymer waveguide. Appl. Phys. Lett. **97**, 121109 (2010)

36. P. Ganguly, J.C. Biswas, S.K. Lahiri, Analysis of Ti:LiNbO$_3$ zero-gap directional coupler for wavelength division multiplexer/demultiplexer. Opt. Commun. **281**, 3269–3274 (2008)

37. E. Armstrong, C.O. Wdyer, Artificial opal photonic crystals and inverse opal structures—fundamentals and applications from optics to energy storage. J. Mater. Chem. C **3**, 6109–6143 (2015)

38. W. Jiang, Study of photonic crystal based waveguide and channel drop filter and localization of light in photonic crystal. Master's thesis, University of Texas, Austin, 2000

39. J.D. Joannopoulos, R.D. Meade, J.N. Winn, *Photonic Crystals—Molding the Flow of Light*, 2nd edn. (Princeton University Press, Princeton, 2008)

40. S. Combrie, G. Lehoucq, A. Junay, S. Malaguti, G. Bellanca, S. Trillo, L. Menager, J.P. Reithmaier, A.D. Rossi, All-optical signal processing at 10 GHz using a photonic crystal molecule. Appl. Phys. Lett. **103**, 193510 (2013)

41. T.F. Krauss, Slow light in photonic crystal waveguides. J. Phys. D Appl. Phys. **40**, 2666–2670 (2007)

42. B. Yang, Y. Zhu, Y. Jiao, L. Yang, Z. Sheng, S. He, D. Dai, Compact arrayed waveguide grating devices based on small SU-8 strip waveguides. J. Lightwave Technol. **29**, 2009–2014 (2011)

43. H. Benisty, P.H. Lalanne, S. Olivier, M. Rattier, C. Weisbuch, C.J.M. Smith, T.F. Krauss, C. Jouanin, D. Cassagne, Finite-depth and intrinsic losses in vertically etched two-dimensional photonic crystals. Opt. Quantum Electron. **34**, 205–215 (2002)

44. L. Cai, H. Han, S. Zhang, H. Hu, K. Wang, Photonic crystal slab fabricated on the platform of lithium niobate-on-insulator. Opt. Lett. **39**, 2094–2096 (2014)

45. G.W. Burr, S. Diziain, M.P. Bernal, The impact of finite-depth cylindrical and conical holes in lithium niobate photonic crystals. Opt. Exp. **16**, 6302–6316 (2008)

46. P. Hamel, P. Grinberg, C. Sauvan, P. Lalanne, A. Baron, A.M. Yacomotti, I. Sagnes, F. Raineri, K. Bencheikh, J.A. Levenson, Coupling light into a slow-light photonic-crystal waveguide from a free-space normally-incident beam. Opt. Exp. **21**, 15144–15154 (2013)

47. J. Shi, M.E. Pollard, C.A. Angeles, R. Chen, J.C. Gates, M.B.D. Charlton, Photonic crystal and quasicrystals providing simultaneous light coupling and beam splitting within a low refractive-index slab waveguide. Sci. Rep. **7**, 1812 (2017)

48. S. Samanta, P. Banerji, P. Ganguly, Design and fabrication of SU-8 polymer based photonic crystal waveguide, in *Frontiers in Optics*, Washington, USA (2017), p. JW3A.70

Chapter 5
Design and Development of Polarization-Independent Power Splitter Using Coupled Silicon Waveguides

5.1 Introduction

The design, fabrication,k and characterization of 1×2 polarization-independent 3-dB power splitter using three-coupled silicon wire/rib waveguides are presented in this chapter. This kind of polarization-independent power splitter was first demonstrated by Ganguly et al. [1] using titanium-indiffused lithium niobate technology. More recently, three-coupled silicon waveguides were proposed in Mach–Zehnder interferometer structure for wavelength division multiplexing (WDM) applications in optical networks on chip [2]. Stegmaier et al. [3] used different configurations of three-coupled waveguides in aluminum nitride-on-insulator platform for power splitting applications in nanoscale integrated optic chip. Use of three-coupled waveguide structure increases the operational bandwidth. In this work, design of the device was accomplished by effective index-based matrix method and coupled mode theory. For fabrication, coupled rib waveguides in silicon on insulator (SOI) platform were chosen and conventional photolithography using chrome mask was used. Device characterization, i.e., measured total insertion loss and optical imbalance between the output ports for TE and TM polarized light indicated the polarization-independent behavior of the splitter.

5.2 Power Splitter Using Coupled Silicon Wire Waveguides

5.2.1 Design and Analysis

Figure 5.1 depicts the schematic of three wire-waveguide power splitter; all the waveguides support single-mode operation at a transmitting wavelength of 1.55 μm. The design of these single-mode wire waveguides was performed using effective index-based matrix method (EIMM) as discussed in Chap. 2. After that the critical

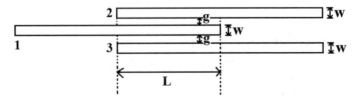

Fig. 5.1 Schematic of three wire-waveguide 1 × 2 power splitter

coupling length (L_c) or coupling coefficient ($\kappa = \pi/2L_c$) of two-coupled single-mode waveguides, which is a key parameter to design devices like power splitter, was computed by using EIMM. In this case, two Lorentzian peaks were obtained in the excitation efficiency versus propagation constant characteristic, one for each of the symmetric and antisymmetric modes ((β_s) and (β_a), respectively). Thus, critical coupling length of the two-coupled waveguide may be obtained from the relation: $L_c = \frac{\pi}{\beta_s - \beta_a}$; however, this L_c value is polarization-dependent and is different for TE and TM modes of the waveguide.

In Fig. 5.1, '1' represents the input at the central waveguide, '2' and '3' represent the outer waveguides; w, L, and g denote the width of the waveguide, length at the coupled region, and separation of the two outer waveguides from the central one, respectively; so, the coupling coefficient is expected to be same in both the outer waveguides. The coupled mode analysis for three-coupled waveguides [4] yields:

$$\begin{bmatrix} A_2(L) \\ A_1(L) \\ A_3(L) \end{bmatrix} = \begin{bmatrix} \frac{1}{2}(1 + \cos\sqrt{2}\kappa L) & \frac{i}{\sqrt{2}}\sin\sqrt{2}\kappa L & -\frac{1}{2}(1 - \cos\sqrt{2}\kappa L) \\ \frac{i}{\sqrt{2}}\sin\sqrt{2}\kappa L & \cos\sqrt{2}\kappa L & \frac{i}{\sqrt{2}}\sin\sqrt{2}\kappa L \\ -\frac{1}{2}(1 - \cos\sqrt{2}\kappa L) & \frac{i}{\sqrt{2}}\sin\sqrt{2}\kappa L & \frac{1}{2}(1 + \cos\sqrt{2}\kappa L) \end{bmatrix} \begin{bmatrix} A_2(0) \\ A_1(0) \\ A_3(0) \end{bmatrix} \quad (5.1)$$

where $A_i(0)$ and $A_i(L)$ are the amplitudes at input and output ends of the three waveguides; $i = 1$ for the central waveguide, and its 2 and 3 for the outer waveguides. From the above relation (5.1), it may be concluded that for $L = L_c/\sqrt{2}$, the total input light energy from the central waveguide 1 will split equally to the outer waveguides (2 and 3) for either TE or TM polarization. Here, $L_c = \pi/2\kappa$ is the critical coupling length between two adjacent coupled waveguides; κ being the coupling coefficient. Now, from the computed L_c values, design of a three-waveguide power splitter for any polarization (TE or TM) can be done easily, provided the coupling length of the device (L) is set at $L_c/\sqrt{2}$. However, there will be an excess loss if the value of the coupling length is not exact, as some light would still be present in the central waveguide 1 which is terminated, though the imbalance of the output ports would remain same. Thus, the imbalance between the outputs is generally polarization-independent for three-coupled waveguides, as opposed to excess loss (Π) which depends on L and hence on TE or TM polarization. The excess loss can be computed (for both TE and TM polarizations) for different coupling length (L) from the following equation:

$$\Pi = -10\log_{10}\left[\{|A_2(L)|^2 + |A_3(L)|^2\}/|A_1(0)|^2\right] \quad (5.2)$$

5.2.2   Computed Results

All the data in this work was obtained using Visual C++ programming language; the refractive indices of Si, SiO$_2$, and air were taken as 3.477, 1.447, and 1.0, respectively, at 1.55 μm wavelength. As found from Chap. 2, the silicon wire waveguide operated under single-mode condition within film thickness 0.0252 μm and 0.2709 μm for TE mode, and 0.1061 μm and 0.3532 μm for TM mode; the single-mode width for TE and TM polarizations ranged in between 0.013–0.290 μm, and 0.010–0.470 μm, respectively. The chosen single-mode silicon layer thickness and width for this work was 0.25 μm. The corresponding guided mode propagation constants of this Si wire waveguide were 9.5193 μm^{-1} and 4.6189 μm^{-1} for TE and TM modes, respectively (as depicted in Figs. 2.8a and 2.9a of Chap. 2).

The excitation efficiency versus propagation constant plot for two straight silicon waveguides with a separation of 0.15 μm for TE mode is depicted in Fig. 5.2. Now, L_c may be computed from: $L_c = \frac{\pi}{\beta_s - \beta_a}$. Figure 5.3 shows the computed data of critical coupling lengths for different separations between the two-coupled silicon wire waveguides for TE and TM polarizations. Since TE mode is more confined compared to TM mode, L_c values of TE mode are larger than the later one. The computed results as obtained using EIMM were compared with commercially available OptiFDTD software, and it was observed that both the results were in good agreement with one another (Fig. 5.3).

The chosen separation between the waveguides was taken as 0.15 μm for the design of three-waveguide polarization-independent power splitter. At this separation, L_c values for TE and TM polarizations were 3.738 μm and 2.507 μm, respectively, as can be seen from Fig. 5.3. Figure 5.4 shows the computed results of excess loss versus coupling length [as obtained using Eq. (5.2)] for both TE and TM modes. It was observed that excess loss is equal to zero for coupling lengths equal to $L_c/\sqrt{2}$

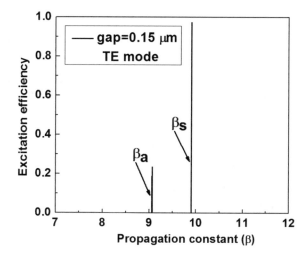

Fig. 5.2 Excitation efficiency versus propagation constant for 0.15 μm gap for TE mode

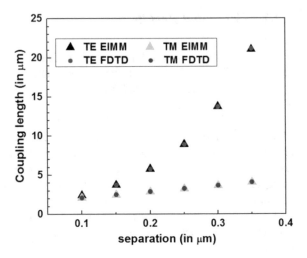

Fig. 5.3 Coupling length versus separation between two Si waveguides (chosen width and thickness = 0.25 μm)

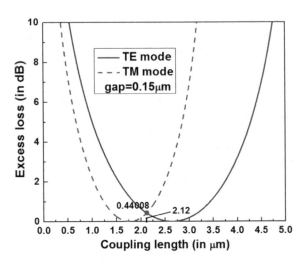

Fig. 5.4 Excess loss for different coupling lengths

(i.e., 2.643 μm for TE mode and 1.773 μm for TM mode). However, at a coupling length of 2.12 μm (the intersection point of excess loss plot of both polarizations), the splitter was polarization-independent, and the excess loss was found to be 0.44 dB irrespective of any polarization.

From the computed data (Fig. 5.5), it is found that the excess loss of the overlapping coupling length (i.e., intersection point of coupling length for TE and TM mode) decreases with the decrease of separation between the waveguides. It may be noted that in this design, the direct coupling between the outer waveguides has been neglected, which is acceptable for the current separation between the waveguides of the coupler.

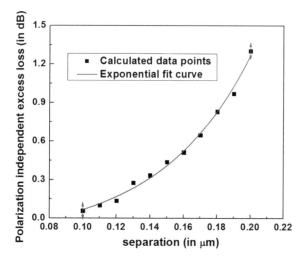

Fig. 5.5 Excess loss of overlapping coupling length for different separations

A study with unequal gaps between waveguides had also been done considering the practical fabrication errors. Figure 5.6 shows the change of excess losses as the difference in gap increases from the desired one of 0.15 μm.

It may be noted that change in excess loss of the device for TE and TM polarization critically depends on the difference of gap between the waveguides. Hence, slight fabrication inequality (≥ 10 nm) of gaps will result into polarization-dependent excess loss for three-waveguide power splitter. Since the fabrication tolerance of electron beam lithography is less than 10 nm, the designed three-waveguide polarization-independent power splitter can be easily fabricated.

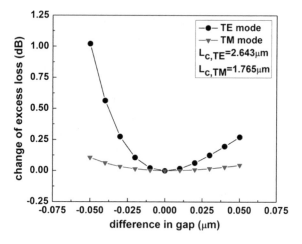

Fig. 5.6 Change of excess loss as difference in gap increases from desired one of 0.15 μm

5.2.3   Discussions

The designed compact power splitter may be fabricated using electron beam lithography and dry etching systems and has higher usable bandwidth around 1.55 μm because of three-coupled waveguides. For this device, one has to use S-type bent waveguides at the outputs. Since Si wire waveguide is a high-contrast waveguide, one can use small bending radii with negligible bending losses for both TE and TM polarization. Details of bending loss computation were discussed in Chap. 2. The power splitter design can be extended for any compact 1×2^N splitter for optical interconnect applications. Fabrication of the designed power splitter requires costly silicon-on-insulator (SOI) substrate with 0.25-μm-thick uniform device layer, which was not available during our experimentation. So, we had redesigned a 1×2 polarization-independent power splitter using silicon rib waveguides on a SOI platform of 5-μm-thick device layer, which was readily available. In the next section, design, fabrication, and characterization of a 1×2 three-waveguide polarization-independent power splitter using silicon rib waveguides are presented.

5.3   Power Splitter Using Coupled Silicon Rib Waveguides

5.3.1   Design and Analysis

Figure 5.7a shows the schematic of our designed coupled silicon rib waveguide-based 1×2 power splitter; all the waveguides are single-mode in nature at 1.55 μm transmitting wavelength. Here, '1' denotes the input which is the center waveguide, while '2' and '3' are the outer waveguides; w_1 is the rib width; g represents separation of the two outer waveguides from the center waveguide; L and L_{bend} indicate the length at the coupled and two-arc S-bend regions, respectively; R is the radius; D being the lateral offset; 'a', 'b,' and 'c' are the transition regions where there is discontinuity in curvatures between straight and/or curved junctions. First of all, single-mode silicon rib waveguide was designed; Fig. 5.7b is the schematic, where w is the rib width and H is the rib thickness, h being the thickness of slab. The design was accomplished using effective index-based matrix method (EIMM); in the first step, the effective index method was used in the depth direction of waveguide. The thickness of slab and rib was made in a way that the effective refractive index of fundamental mode of rib ($n_{\text{eff},0}$) was always greater than the effective refractive index of fundamental mode of slab ($n_{\text{eff,slab}}$), and all other higher order modes in the core region were less than $n_{\text{eff,slab}}$; thus, the higher order vertical modes in the rib region coupled with the fundamental slab mode of the rib-side regions. Next, transfer matrix method was applied to this resulted lateral refractive index profile.

Then with chosen single-mode parameters, coupling between three-coupled straight rib waveguides were analyzed, for which at first coupling between two straight adjacent waveguides was considered by using EIMM. The lateral effective

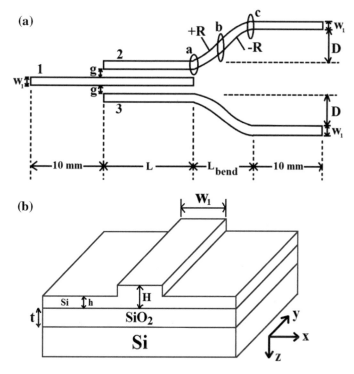

Fig. 5.7 Schematic of **a** coupled silicon rib waveguide-based 1×2 power splitter, **b** silicon rib waveguide (reprinted with permission from [5]. ©2018 IOP Publishing Ltd)

index profile of these two straight coupled waveguides was computed by effective index method; thereafter, this was used in matrix method from which we obtained symmetric (β_s) and antisymmetric (β_a) propagation constant values (from excitation efficiency versus propagation constant plot). As discussed in previous section, since $L_c = \frac{\pi}{\beta_s - \beta_a}$, critical coupling length (L_c) can be calculated; also, excess loss can be found from Eq. (5.1) proceeding similarly. Next, double-arc S-bend waveguide was modeled using two constant radii of curvature R, which can be calculated by [6]: $R = \pm \frac{L_{bend}^2}{4D}\left(1 + \frac{D^2}{L_{bend}^2}\right)$. Here, D is the lateral offset between the two parallel waveguides and L_{bend} is the transition length in the longitudinal direction, which has both bending and transition losses. Pure bending loss was computed by conformal mapping technique [7] along with transfer matrix method [8, 9]. The effective index profile of the bend waveguide was first converted into equivalent straight one by conformal mapping technique, and transfer matrix method was applied on this profile from where we obtained Lorentzian-shaped resonant peak of propagating constant of the waveguide. The value of propagation constant where this peak appeared represents the real part of propagation constant, whereas full-width-at-half-maxima (Γ) represents twice the imaginary part of it. After that, bending loss (BL) was calculated from the relation: BL (in dB/unit length) $= 4.34(\Gamma)$. Thus, pure bending loss for length L_{bend} of a bent S-type waveguide with constant radius of curvature (R) can be found from $BL = 4.34(\Gamma)L_{\text{bend}}$. The transition region between the junctions of straight and/or arc bends will also have some loss, which was computed using the following formula [6]:

$$T = -4.34 \ln\left(1 - \frac{\pi^2 \partial_m^2}{4w_1^2}\right)^2 \tag{5.3}$$

Here, T is the transition loss; ∂_m is the modal offset between two arc bends, which is very small in all practical cases; w_1 being the waveguide width. If R_1 and R_2 are the radii of curvature of two waveguide bends, then the modal offset may be shown as $\partial_m = \left(\frac{V_1^2 a_x^4}{\rho^2 \Lambda}\right)\left(\frac{1}{R_1} - \frac{1}{R_2}\right)$, provided the fundamental mode of the waveguide is well approximated by a Gaussian distribution of the form: $E(x) = E_0 \exp\left(-\frac{x^2}{2a_x^2}\right)$. Here $\Lambda = \frac{n_{\text{rib}}^2 - n_{\text{slab}}^2}{2n_{\text{rib}}^2}$, ρ is the half-width of waveguide, $V_1 = \frac{2\pi}{\lambda}\rho n_{\text{rib}}\sqrt{2\Lambda}$, n_{rib} and n_{slab} are the respective refractive indices of core rib and slab; λ is the wavelength; $a_x = A_x\rho$ is the spot size and is computed solving eigenvalue equation of form: $\exp\left(\frac{1}{x^2}\right) = 2A_x\left(\frac{V_1^2}{\sqrt{\pi}}\right)$. We had chosen L_{bend} and D in such a way so that total bending loss of S-type silicon rib waveguides for TE and TM polarizations was nearly equal and negligibly small.

5.3.2 Computed Results

It was found by using effective index-based matrix method that within the width 0.9–6.2 μm for TE mode and 0.5–5.4 μm for TM mode, the rib waveguide was single-mode in nature. The chosen rib and slab thicknesses were 5 μm and 3.5 μm, respectively, which supported single-mode condition in the depth direction for both TE and TM modes. On the other hand, in lateral direction, 5 μm rib width was chosen. The plot of coupling length for different separation between two adjacent coupled straight waveguides for both polarizations is shown in Fig. 5.8. We had chosen 5 μm separation between the waveguides, so that direct coupling between the outer waveguides might be neglected. Now, at a coupling length of 4.31 mm (for TE mode) and 5.64 mm (TM mode), i.e., at $L = L_c/\sqrt{2}$, the power from central waveguide 1 splits equally to the outer waveguides 2 and 3 (as can be seen from Fig. 5.9), which is obvious from Eq. (5.1). Figure 5.10 shows the excess loss value with respect to coupling length, the gap being 5 μm, where it can be seen that at $L = L_c/\sqrt{2}$, the excess loss value is zero for both TE and TM modes. Figure 5.10 also illustrates that at overlapping coupling length 4.89 mm, the excess loss value is 0.191 dB, which is polarization-independent.

Fig. 5.8 Coupling length for
different separation between
two adjacent straight
waveguides

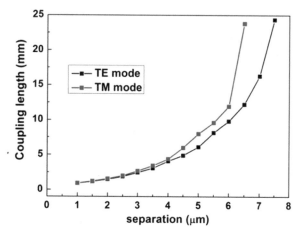

Fig. 5.9 Power distribution
from center waveguide 1 to
outer arms 2 and 3 **a** TE
mode, **b** TM mode

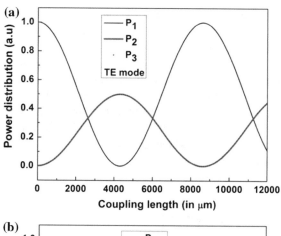

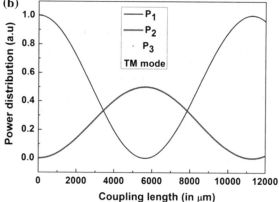

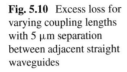

Fig. 5.10 Excess loss for varying coupling lengths with 5 μm separation between adjacent straight waveguides

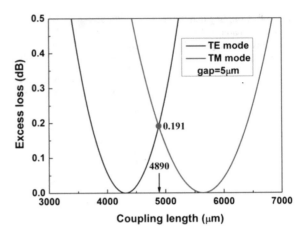

For the double-arc S-bend curve, the chosen L_{bend} and D were 10 mm and 100 μm, respectively, which yield the total bending loss to be negligibly small for both TE and TM modes, where R was taken as 250 mm.

Figure 5.11 depicts the decrease of bending losses with increase in radii of curvature for both polarizations. The transition loss between straight and bend junction, i.e., region 'a' and 'c' of Fig. 5.7a was 0.000933 dB for TE mode and 0.00072 dB for TM mode, whereas the transition loss for region 'b,' i.e., junction between two bend waveguides was 0.0149 dB and 0.0115 dB for TE and TM mode respectively. Thus, the total transition loss for TE mode was 0.0167 dB and for TM mode the loss value was 0.0129 dB.

5.3.3   Fabrication

Fabrication was done by optical lithography using 0.1-μm-thick chrome mask which was coated with 0.5 μm AZ 1500 positive photoresist. Pattern on this chrome mask plate was made by laser direct writing system with UV laser source of 405 nm. The calibrated gain and corresponding bias were 21.8 and 102 mJ/cm², respectively. Development was done for 30 s in a developer solution and then dipped in chrome etchant for 2 min to get the pattern of power splitter and straight waveguides. Finally, the photoresist was stripped by dipping the mask plate in microstrip 3001 for 1 min and next 1 min in acetone. Next, an SOI wafer was taken and was cleaned by 1:1 volume ratio of hydrogen peroxide and concentrated sulfuric acid for 30 min. After drying with nitrogen jet, it was undergone with general heating for 30 min and then placed in a thermal deposition system (Milman, India) in order to obtain a layer of aluminum on top of silicon surface. With 2.7 g/cm³ density and acoustic impedance of 17.10 g/cm² s × 10⁵, aluminum thickness of ~230 nm was deposited (as observed in thickness monitor as well as measured from Dektak surface profiler). Then positive

Fig. 5.11 Bending loss for
varying radii of curvature
a TE mode, **b** TM mode

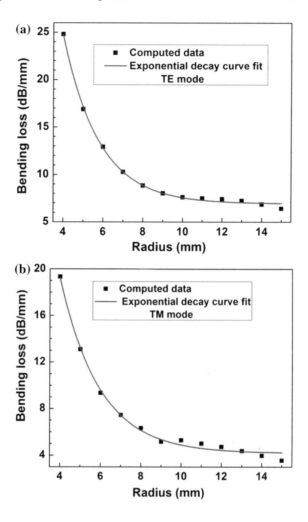

photoresist (HPR 504) was spun at 500 rpm for 10 s and next 20 s at 3000 rpm using a
spin coater. Pre-bake of sample at 90 °C was done for 30 min for solvent evaporation
and then it was exposed to UV light of 365 nm laser source using MJB 3 mask aligner
for 7.5 s. With HPRD 429 positive resist developer, development was done carefully
for 25 s and was rinsed with deionized (DI) water before post-exposure baking at
120 °C for 30 min. After that, the surrounding aluminum layer (not covered by
the resist) was removed using aluminum etchant at an etch rate of 0.33 nm/min.
Next, we proceeded with reactive ion etching (RIE) of silicon, where an etch depth
of ~1.8 μm of silicon from areas not protected by the aluminum hard mask was
obtained. Parameters like pressure, gas flow rate, power, and process duration were
calibrated. Figure 5.12 shows the plot for different etch depths with time; Table 5.1
is the optimized RIE process parameters having roughness ~30 nm.

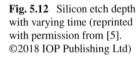

Fig. 5.12 Silicon etch depth with varying time (reprinted with permission from [5]. ©2018 IOP Publishing Ltd)

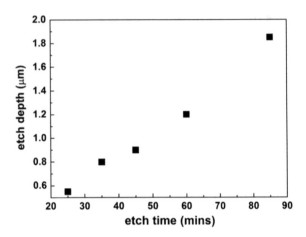

Table 5.1 Optimized RIE process parameters

Pressure (mTorr)	Flow rate (sccm)		Power (Watt)	Time (mins)
	SF$_6$	O$_2$		
100	5	1	50	85

Reprinted with permission from [5]. ©2018 IOP Publishing Ltd

5.3.4 Characterization

The sample was cleaved and inspected under an optical microscope and scanning electron microscope; Fig. 5.13a shows photomicrograph of edge of fabricated silicon rib waveguide. It may be noted that our dry etching process yields around 60° etching profile and a fairly smooth end-face of the fabricated waveguide was obtained after cleaving. Figure 5.13b is the scanning electron microscope view of three-coupled waveguide region. Figure 5.14a, b depict the scanning electron microscope view of straight to three-coupled waveguide region and three-coupled waveguide to arm split region, respectively. Table 5.2 is the comparative results between the designed and fabricated data of waveguide width and gap between adjacent waveguides. We can see that there are slight differences between the fabricated and designed values, which occurred due to fabrication error during lithography and etching processes.

Mode image of the fabricated single-mode silicon rib waveguide was observed and recorded at 1.55 μm transmitting wavelength. Mode image of this large core silicon waveguide is shown in Fig. 5.15. The total insertion loss of this silicon rib waveguide of length 2.7 cm was measured using an optical arrangement as shown in Fig. 3.2 and was found to be 8.4 dB and 10.0 dB for TE and TM mode, respectively. Characterization of the fabricated power splitter was also carried out using the same optical setup.

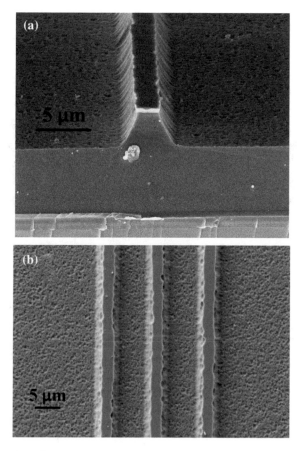

Fig. 5.13 SEM view of **a** rib waveguide edge, **b** three-coupled rib waveguide region (reprinted with permission from [5]. ©2018 IOP Publishing Ltd)

The obtained total insertion loss of the power splitter for TE mode was 11.43 dB and that of TM mode was 11.8 dB, which shows that total insertion loss is fairly polarization-independent. As the fabricated gaps of the coupler between the outer arms from the central input waveguide were not exactly same, we observed an imbalance in the power splitting ratio. The measured imbalance as calculated from the relation: $-10 \log(P_2/P_3)$ for TE mode was 0.23 dB, while the value for TM mode was 0.82 dB, where P_2 and P_3 being the power outputs of outer arms 2 and 3, respectively.

The obtained total insertion losses of rib waveguide in SOI platform were compared with other published works, which is shown in Table 5.3. It may be noted that waveguide loss is on the higher side compared to other reported results, which indicates that etching process has to be optimized more to reduce the side-wall roughness of waveguide. Also, a comparative study (Table 5.4) was made between three-waveguide power splitter with other available splitter configurations. It is clear

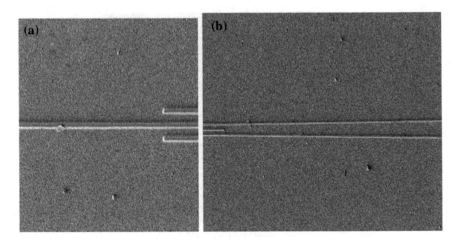

Fig. 5.14 SEM view of **a** straight to three-coupled waveguide region, **b** three-coupled waveguide to arm split region (reprinted with permission from [5]. ©2018 IOP Publishing Ltd)

Table 5.2 Comparison of designed and fabricated parameters of the power splitter

	Slab height 'h' (μm)	Rib height 'H' (μm)	Ridge height 'H–h' (μm)	Width 'w' (μm)		Gap 'g' (μm)	
				Input	Outer arms	Between input and outer arm 2	Between input and outer arm 3
Designed data	3.50	5.0	1.50	5.0	5.0, 5.0	5.0	5.0
Fabricated data	3.15	5.0	1.85	4.7	4.8, 4.7	5.2	5.6

Reprinted with permission from [5]. ©2018 IOP Publishing Ltd

Fig. 5.15 Mode image of single-mode silicon rib waveguide for TE mode (reprinted with permission from [5]. ©2018 IOP Publishing Ltd)

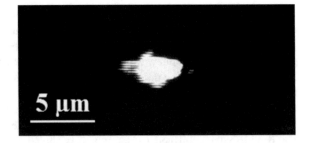

5 μm

from this table that apart from being fairly polarization-independent, total insertion loss and imbalance between output ports of our fabricated device and other reported results are of same order.

Table 5.3 Comparison of measured total insertion loss of silicon rib waveguide with other published works at 1.55 μm wavelength

Slab and rib height ratio (h/H)	Rib width (μm)	Length of waveguide (cm)	Insertion loss (dB)		References
			TE	TM	
0.70	5.0	1.0	~11.5	~12.0	[10]
0.44	6.7	2.7	~3.8 (unpolarized)		[11]
0.64	5.0	2.7	8.4	10.0	[5] Our work

Reprinted with permission from [5]. ©2018 IOP Publishing Ltd

Table 5.4 Comparison of characterized results of silicon power splitter with other published configurations

1 × 2 power splitter configuration	Operating wavelength (μm)	Insertion loss (dB)		Imbalance (dB)		References
		TE	TM	TE	TM	
Multi-mode interference-based (using Si rib waveguides)	1.55	10.5	–	0.2	–	[12]
Multi-mode interference-based (using Si wire waveguides)	1.55	~27.0	–	~0.5	–	[13]
Arc-shaped Y-splitter (using Si wire waveguides)	1.55	11.0	–	0.07	–	[14]
Three-coupled waveguide based (using Si rib waveguides)	1.55	11.43	11.80	0.23	0.82	Our work [5]

Reprinted with permission from [5]. ©2018 IOP Publishing Ltd

5.4   Conclusions

This chapter presents development of a three-waveguide polarization-independent power splitter using SOI platform. Initially, for compact devices, the design was performed using silicon wire waveguides of 0.25 μm × 0.25 μm cross-sectional area. The design can be extended for any 1×2^N 3-dB power splitter which will be very useful for optical interconnects and fiber-optic communication network. Apart from polarization-independent nature, use of three-coupled waveguides reduces the power imbalance between the output ports and also increases the operational optical bandwidth compared to power splitters made of two-coupled waveguides [15–17]. Due to non-availability of SOI wafer of 0.25-μm-thick device layer during our experimentation, we also designed a 1×2 three-waveguide polarization-independent power splitter using large core silicon rib waveguides. For proof-of-concept, the component was fabricated using a readily available SOI substrate of 5-μm-thick device layer and characterized at 1.55 μm wavelength of light. The fabricated device has insertion loss of 11.43 dB for TE polarization and 11.80 dB for TM mode, which indicates its polarization-independent behavior. The power splitting shows an imbalance of 0.23 dB for TE mode and 0.82 dB for TM mode, which is due to fabrication error of separation of outer arms from the central input waveguide.

References

1. P. Ganguly, J.C. Biswas, S. Das, S.K. Lahiri, A three-waveguide polarization independent power splitter on lithium niobate substrate. Opt. Commun. **168**, 349–354 (1999)
2. G. Calo, A. D'Orazio, V. Petruzzelli, Broadband Mach-Zehnder switch for photonic networks on chip. J. Lightwave Technol. **30**, 944–952 (2012)
3. M. Stegmaier, W.H.P. Pernice, Broadband directional coupling in aluminum nitride nanophotonic circuits. Opt. Exp. **21**, 7304–7315 (2013)
4. H. Ogiwara, Optical waveguide 3×3 switch: theory of tuning and control. Appl. Opt. **18**, 510–515 (1979)
5. S. Samanta, P.K. Dey, P. Banerji, P. Ganguly, A 1×2 polarization-independent power splitter using three-coupled silicon rib waveguides. J. Opt. **20**, 095801 (2018)
6. P. Ganguly, J.C. Biswas, S.K. Lahiri, Modelling of titanium indiffused lithium niobate channel waveguide bends: a matrix approach. Opt. Commun. **155**, 125–134 (1998)
7. M. Heiblum, Analysis of curved optical waveguides by conformal transformation. IEEE J. Quantum Electron. **11**, 75–83 (1975)
8. A. Ghatak, K. Thyagarajan, M.R. Shenoy, Numerical analysis of planar optical waveguides using matrix approach. J. Lightwave Technol. **5**, 660–667 (1987)
9. M.R. Shenoy, K. Thyagarajan, A. Ghatak, Numerical analysis of optical fibers using matrix approach. J. Lightwave Technol. **6**, 1285–1291 (1988)
10. P.K. Dey, P. Ganguly, A technical report on fabrication of SU-8 optical waveguides. J. Opt. **43**, 79–83 (2014)
11. K. Solehmainen, T. Aalto, J. Dekker, M. Kapulainen, M. Harjanne, K. Kukli, P. Heimala, K. Kolari, M. Leskela, Dry-etched silicon-on-insulator waveguides with low propagation and fiber-coupling losses. J. Lightwave Technol. **23**, 3875–3880 (2005)
12. J.S. Xia, J.Z. Yu, Z.C. Fan, Z.T. Wang, S.W. Chen, Multimode interference 3-dB coupler in silicon-on-insulator based on silicon rib waveguides with trapezoidal cross section. Chin. Phys. Lett. **21**, 104–106 (2004)
13. Z. Jingtao, Z. Huihui, L. Xinyu, Design and fabrication of a compact multimode interference splitter with silicon photonic nanowires. Chin. Opt. Lett. **7**, 1041–1044 (2009)
14. S.H. Tao, Q. Fang, J.F. Song, M.B. Yu, G.Q. Lo, D.L. Kwong, Cascade wide-angle Y-junction 1×16 optical power splitter based on silicon wire waveguides on silicon-on-insulator. Opt. Exp. **16**, 21456–21461 (2008)
15. Y. Quan, P.D. Han, Q.J. Ran, F.P. Zeng, L.P. Gao, C.H. Zhao, A photonic wire-based directional coupler based on SOI. Opt. Commun. **281**, 3105–3110 (2008)
16. H. Fukuda, K. Yamada, T. Tsuchizawa, T. Watanabe, H. Shinojima, S.I. Itabashi, Ultrasmall polarization splitter based on silicon wire waveguides. Opt. Exp. **14**, 12401–12408 (2006)
17. J. Chee, S. Zhu, G.Q. Lo, CMOS compatible polarization splitter using hybrid plasmonic waveguide. Opt. Exp. **20**, 25345–25355 (2012)

Chapter 6
Conclusions and Future Scope of Work

6.1 Summary

The monograph deals with research topic related to polymer and silicon waveguides, and components for integrated optic applications. Effective index-based matrix method (EIMM) was extended to step-index silicon (Si) and polymer (SU-8) waveguides. SU-8 wire waveguides were fabricated by maskless continuous-wave direct laser writing technique at 375 nm writing wavelength; detailed characterizations were made using in-house laboratory facilities to validate the computed results of EIMM. Design, fabrication, and characterization of single-mode SU-8 waveguide, directional coupler, and micro-ring resonator were carried out, where fabrication was done using conventional I-line photolithography with chromium mask. A feasibility study on photonic crystal structure fabricated on straight SU-8 wire waveguide was also performed both theoretically and experimentally. Single-mode silicon rib waveguide was fabricated using SOI substrate, and a three-coupled-waveguide polarization-independent power splitter was designed and demonstrated using these waveguides. A brief review on SU-8 and Si waveguides, micro-ring resonators, photonic crystal waveguides, and power splitters was also made as the background work of the book.

6.2 Contributions and Achievements

The principal contributions of the monograph may be summarized as follows:

1. The review on silicon and SU-8 optical waveguides, micro-ring resonators, photonic crystal structure, and power splitter, presented in Chap. 1 may be useful for researchers working in this field. The advantages of indigenously developed design tool, EIMM, over mostly used numerical techniques for integrated optic applications are also included.

2. Effective index-based matrix method (EIMM) has been extended for step-index wire and rib waveguides in Chap. 2. Silicon and SU-8 polymer waveguides on silicon substrate were considered for the purpose. EIMM is a two-step process; in the first step, effective index method was used for vertical refractive index profile of waveguide, and then for the resulted lateral index profile, a transfer matrix method was applied. The method was successfully applied in designing single-mode waveguides, estimating bending loss of bent waveguides, and computing lateral mode profiles for the guided modes. EIMM was found to be less computation-intensive than commercially available numerical softwares, such as OptiFDTD.

3. Report of air- and PDMS-cladded SU-8 wire waveguides fabricated using direct laser writing at 375 nm wavelength is given in Chap. 3. The fabricated waveguides were characterized in detail, and some of the characterization results, such as mode index, lateral mode profile, and refractive index profiles were successfully validated with the theoretically predicted data using EIMM. Measured propagation losses of these SU-8 waveguides at 1550 nm transmitting wavelength with air- and PDMS-cladding were 0.51 dB/mm and 0.30 dB/mm, respectively, which are of same order of magnitude comparing with other previously reported data. In the end, a feasibility study to fabricate SU-8 waveguide structures by focused ion beam (FIB) lithography was done and the method was found suitable in making precise modifications in micro and nanoscale photonic waveguide structures, instead of long waveguides.

4. Design and development of wire waveguide structures, viz. directional coupler and micro-ring resonator (MRR), using SU-8 polymer are discussed in Chap. 4. Fabrication of these structures was done using conventional I-line photolithography using chrome mask instead of laser direct writing technique, to achieve better precision and control over fabricated waveguide structures. The work indicated the possibility of using optical lithography instead of using costly electron beam lithography system to fabricate MRR. Minimum separation between two coupled waveguides of directional coupler was obtained as 0.57 μm by optical lithography. The fabricated MRR of 15 μm radius was characterized using semiconductor laser diode and a calibrated monochromator. Characterization results showed that it could be used as a bandpass filter around 1565 nm wavelength of light with a 3-dB bandwidth of 5.36 nm for TE polarization. Some theoretical studies of photonic crystal structures on SU-8 wire waveguide were also conducted, and accordingly, a fabrication attempt of the structure was also made using FIB lithography.

5. Design and demonstration of a 1 × 2 polarization-independent 3-dB power splitter using three-coupled silicon wire/rib waveguides are presented in Chap. 5. The component was fabricated using a readily available SOI substrate of 5-μm-thick device layer and characterized at 1.55 μm wavelength of light. Insertion loss of 11.43 dB for TE polarization and 11.80 dB for TM mode indicated its polarization-independent behavior. The power splitting showed an imbalance of 0.23 dB for TE mode and 0.82 dB for TM mode, which was due to fabrication error of separation of outer arms from the central input waveguide.

6.3 Limitations of the Present Work and Scopes of Future Research

The limitations of the present work and some comments on the future scope of research in this field are highlighted below:

1. A brief review on Si and SU-8 optical waveguides, MRR and power splitter has been made in Chap. 1. Obviously, this is not a complete up-to-date review on the subject, and there may be some omissions of important references as well due to limitations of library facilities. Any such omission is regretted.

2. Effective index-based matrix method (EIMM) was used to design Si and SU-8 waveguides on oxidized Si substrate. It has been shown in Chap. 2 (Fig. 2.7) that the computation technique is not that accurate near the cut-off region of the guided modes in-depth direction. The deviations between results of effective index method and 2D-FDTD, applied in-depth direction, increase with decrease of refractive index contrast. The problem with the effective index method is that its accuracy depends very much on the waveguide structure. The effective index method for rectangular structure, such as Si and SU-8 wire waveguides, can be improved further to provide much more accurate results even near cut-off, without sacrificing the computational efficiency. Such a method is called the effective index method (EIM) with built-in perturbation correction [1–3]. The computation process presented in the book may be made more accurate by taking account of this improved EIM.

3. Propagation loss and fiber–waveguide coupling loss of fabricated air-cladded and PDMS-cladded SU-8 waveguide were measured and given in Chap. 3. These values can be even lowered by proper optimization of fabrication parameters as well as better edge preparation of waveguides. Moreover, the trapezoidal cross-sectional profile of waveguides, as shown in Fig. 3.1c, may be made rectangular in shape by optimizing SU-8 processing parameters and using special fused silica objective lens during writing. In that case, comparison between computed and experimental results would be more realistic.

4. The obtained Q-value of fabricated SU-8 MRR as described in Chap. 4 is quite low (~292). There may be three reasons behind it: (i) the bus and ring waveguides of width 3.5 μm are not strictly single-mode, (ii) resolution of monochromator used during measurement is low (~0.6 nm), and (iii) bending loss of SU-8 micro-ring of 15 μm radius and propagation loss of waveguides are high. To improve the Q-value of MRR described in this chapter, one may use electron beam lithography instead of optical lithography to fabricate strictly single-moded structure of lower waveguide width. For spectral characterization, tunable laser source (TLS)—optical spectrum analyzer/photodetector assembly may be used for accurate measurement. Finally, radius of SU-8 MRR may be increased to reduce the bending loss. This will also increase total propagation loss other than bending loss. Hence, to increase Q-value of MRR, optimized structure with lower propagation loss is required, which may be investigated further.

5. The photonic crystal structure on SU-8 wire waveguide fabricated by FIB lithography was discussed at the end of Chap. 4. The structure may be useful for input/output light coupling into the waveguide as well as for polarization-independent bandpass optical filter. The optimization of fabrication process and detailed optical characterization of the photonic crystal waveguide were not performed yet. A detailed study on fabrication and optical characterization of these waveguides can be undertaken in future.

6. Chapter 5 presents design and demonstration of three-waveguide polarization-independent power splitter using Si rib waveguides in SOI platform. In theoretical analysis of three coupled-waveguides, direct coupling between outer waveguides was not considered. This is valid only for cases where the separation between outer guides is sufficiently large. For more compact structure, however, the direct coupling is to be considered.

7. The design and fabrication of 1×2 power splitter may also be extended to study polarization-independent 1×3 power splitters. It may be noted from Fig. 5.9 that a uniform 1×3 power splitter requires even lower coupling length than that of 1×2 power splitters.

8. Dry etching process of Si is to be optimized to reduce the waveguide propagation loss of fabricated rib waveguides. Apart from trapezoidal cross section, the side wall roughness of our first fabricated rib waveguides was quite high (Fig. 5.13), which made the waveguide lossy.

References

1. K.S. Chiang, K.M. Lo, K.S. Kwok, Effective-index method with built-in perturbation correction for integrated optical waveguides. J. Lightwave Technol. **14**, 223–228 (1996)
2. K.S. Chiang, C.H. Kwan, K.M. Lo, Effective-index method with built-in perturbation correction for the vector modes of rectangular-core optical waveguides. J. Lightwave Technol. **17**, 716–722 (1999)
3. C.H. Kwan, K.S. Chiang, Study of polarization-dependent coupling in optical waveguide directional couplers by the effective-index method with built-in perturbation correction. J. Lightwave Technol. **20**, 1018–1026 (2002)